Catalysts for Fine Chemical Synthesis

Volume 1

Catalysts for Fine Chemical Synthesis

Series Editors

Stan M Roberts, Ivan V Kozhevnikov and **Eric Derouane**
University of Liverpool, UK

Forthcoming Volumes

Catalysts for Fine Chemical Synthesis Volume 2
Catalysis by Polyoxometalates
Ivan V Kozhevnikov
University of Liverpool, UK

ISBN 0 471 62381 4

Catalysts for Fine Chemical Synthesis Volume 3
Edited by Eric Derouane
University of Liverpool, UK

ISBN 0 471 49054 7

Catalysts for Fine Chemical Synthesis

Volume 1

Hydrolysis, Oxidation and Reduction

Edited by

Stan M Roberts and **Geraldine Poignant**
University of Liverpool, UK

JOHN WILEY & SONS, LTD

Other Wiley Editorial Offices

John Wiley & Sons Inc., 111 River Street,
Hoboken, NJ 07030, USA

Jossey-Bass, 989 Market Street,
San Francisco, CA 94103-1741, USA

Wiley-VCH Verlag GmbH,
Boschstr. 12, D-69469 Weinheim, Germany

John Wiley & Sons Australia Ltd, 33 Park Road, Milton,
Queensland 4064, Australia

John Wiley & Sons (Asia) Pte Ltd, 2 Clementi Loop # 02-01,
Jin Xing Distripark, Singapore 129809

John Wiley & Sons Canada Ltd, 22 Worcester Road,
Etobicoke, Ontario, Canada M9W 1L1

Library of Congress Cataloging-in-Publication Data

Hydrolysis, oxidation, and reduction / edited by Stan M. Roberts and Geraldine Poignant.
    p. cm—(Catalysts for fine chemical synthesis; v. 1)
    Includes bibliographical references and index.
    ISBN 0-471-49850-5 (acid-free paper)
       1. Enzymes—Biotechnology. 2. Organic compounds—Synthesis. 3. Hydrolysis.
    4. Oxidation-reduction reaction. I. Roberts, Stanley M. II. Poignant, Geraldine. III. Series.

    TP248.65.E59 H98 2002
    660'.28443—dc21

                                                                              2002072357

British Library Cataloguing in Publication Data
A catalogue record for this book is available from the British Library

ISBN 0 471 98123 0

Typeset in 10/12pt Times by Kolam Information Services Pvt Ltd, Pondicherry, India.
Printed and bound in Great Britain by Antony Rowe Ltd, Chippenham, Wiltshire.
This book is printed on acid-free paper responsibly manufactured from sustainable forestry in which at
least two trees are planted for each one used for paper production.

Contents

Catalysts for Fine Chemical Synthesis
Series Preface

During the early-to-mid 1990s we published a wide range of protocols, detailing the use of biotransformations in synthetic organic chemistry. The procedures were first published in the form of a loose-leaf laboratory manual and, recently, all the protocols have been collected together and published in book form (*Preparative Biotransformations*, Wiley-VCH, 1999).

Over the past few years the employment of enzymes and whole cells to carry out selected organic reactions has become much more commonplace. Very few research groups would now have any reservations about using commercially available biocatalysts such as lipases. Biotransformations have become accepted as powerful methodologies in synthetic organic chemistry.

Perhaps less clear to a newcomer to a particular area of chemistry is *when* to use biocatalysis as a key step in a synthesis, and when it is better to use one of the alternative non-natural catalysts that may be available. Therefore we set out to extend the objective of *Preparative Biotransformations*, so as to cover the whole panoply of catalytic methods available to the synthetic chemist, incorporating biocatalytic procedures where appropriate.

In keeping with the earlier format we aim to provide the readership with sufficient practical details for the preparation and successful use of the relevant catalyst. Coupled with these specific examples, a selection of the products that may be obtained by a particular technology will be reviewed.

In the different volumes of this new series we will feature catalysts for oxidation and reduction reactions, hydrolysis protocols and catalytic systems for carbon–carbon bond formation *inter alia*. Many of the catalysts featured will be chiral, given the present day interest in the preparation of single-enantiomer fine chemicals. When appropriate, a catalyst type that is capable of a wide range of transformations will be featured. In these volumes the amount of practical data that is described will be proportionately less, and attention will be focused on the past uses of the system and its future potential.

Newcomers to a particular area of catalysis may use these volumes to validate their techniques, and, when a choice of methods is available, use the background information better to delineate the optimum strategy to try to accomplish a previously unknown conversion.

S.M. ROBERTS
I. KOZHEVNIKOV
E. DEROUANE
LIVERPOOL, 2002

Preface for Volume 1: Hydrolysis, Oxidation and Reduction

A REVIEW OF NATURAL AND NON-NATURAL CATALYSTS IN SYNTHETIC ORGANIC CHEMISTRY: PRACTICAL TIPS FOR SOME IMPORTANT OXIDATION AND REDUCTION REACTIONS

In this volume we indicate some of the different natural and non-natural catalysts for hydrolysis, oxidation, reduction and carbon–carbon bond forming reactions leading to optically active products. Literature references are given to assist the reader to pertinent reviews. The list of references is not in the least comprehensive and is meant to be an indicator rather than an exhaustive compilation. It includes references up to mid-1999 together with a handful of more recent reports.

The later sections of the book deal with the actual laboratory use of catalysts for asymmetric reduction and oxidation reactions. Most of the protocols describe non-natural catalysts principally because many of the corresponding biological procedures were featured in the sister volume *Preparative Biotransformations*. As in this earlier book, we have spelt out the procedures in great detail, giving where necessary, helpful tips and, where appropriate, clear warnings of toxicity, fire hazards, etc.

Many of the procedures have been validated in the Liverpool laboratories (by GP). Other protocols were kindly submitted by colleagues from the USA, Japan, the UK and mainland Europe. The names of the contributors are given at the start of the corresponding protocol. These descriptions of the recipes also contain references to the literature. In these cases the references point the reader to the more practical aspects of the topic and are meant to complement rather than repeat the references given in the first, overview chapter.

Some of the practicals describe the use of similar catalysts and/or catalysts that accomplish the same task. This has been done purposely to try to get the best match between the substrate described and the one being considered by an interested reader. Moreover when catalysts can be compared, this has been done. Sometimes a guide is given as to what *we* found to be the most useful system in our hands. In this context, it is important to note that, except for polyleucine-catalysed oxidations and the use of a bicyclic bisphosphinite for asymmetric hydrogenation, the Liverpool group had no previous experience in using the catalysts described herein; we approached the experiments carried out in Liverpool as newcomers in the field.

Thus for the first volume in this series we have performed a selection of oxidation and reduction reactions, arguably some of the most important transformations of these two types, mainly employing non-natural catalysts. In other volumes of this work other catalysts for oxidation and reduction will be featured and, of equal importance, the use of preferred catalysts for carbon–carbon bond formation will be described. In the first phase, therefore, this series will seek to explore the 'pros and cons' of using many, if not most, well-documented catalysts and we will endeavour to report our findings in a non-partisan manner.

We truly hope these procedures will be really valuable for fellow chemists trying out a new catalyst system for the first time. Feedback and further hints and tips would be most welcome.

G. POIGNANT
S.M. ROBERTS
LIVERPOOL, 2002

Abbreviations

Ac	acetyl
Ar	aryl
b.p.	boiling point
BSA	*N,O*-bis-(trimethylsilyl)-acetamide
Bu	butyl
cat	catalyst
CLAMPS	cross-linked aminomethylpolystyrene
DBU	1,8-diazabicyclo[5.4.0]undec–7–ene
DEPT	diethyl tartrate
DIPT	diisopropyl tartrate
DMAP	4-dimethylaminopyridine
DMM	dimethoxymethane
DMSO	dimethyl sulfoxide
EDTA	ethylenediaminetetraacetic acid
ee	enantiomeric excess
eq	equivalent
Et	ethyl
GC	gas chromatography
HPLC	high pressure liquid chromatography
ID	internal diameter
IR	infrared (spectroscopy)
L	ligand
lit.	literature
M	metal
m.p.	melting point
MCPBA⎫ *m*-CPBA⎭	*meta*-chloroperbenzoic acid
Me	methyl
MTPA	methoxy-α-(trifluoromethyl)phenylacetyl
NMR	nuclear magnetic resonance
Ph	phenyl
Pr	propyl
psi	pounds per square inch
r.p.m.	rotation per minutes
R_f	retention factor
R_t	retention time

TBHP	*tert*-butyl hydroperoxide
THF	tetrahydrofuran
TLC	thin layer chromatography
TMS	tetramethylsilane
UHP	urea–hydrogen peroxide
UV	ultraviolet
v:v	volume per unit volume

Part I   Review

1 The Integration of Biotransformations into the Catalyst Portfolio

CONTENTS

The science of biotransformations has been investigated since the days of Pasteur[1]. However, progress in the use of enzymes and whole cells in synthetic organic chemistry was relatively slow until the 1950s, when the use of microorganisms to modify the steroid nucleus was studied in industry and academic laboratories[2]. Thus conversions such as the transformation of 17α-acetoxy-11-deoxycortisol into cortisol (hydrocortisone) (1), using the microorganism

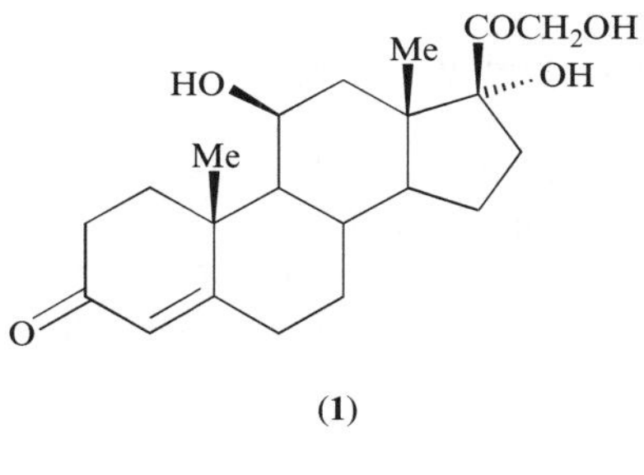

(1)

Curvularia lunata to introduce the 11β-hydroxy group directly, helped to revive interest in the application of biological catalysis to problems in synthetic organic chemistry. The momentum was continued by Charles Sih, J. Bryan Jones, George Whitesides and others, until, by the mid-1980s, biocatalysis was being accepted as a powerful method, especially for the production of optically active products[3]. At this time the whole field was given another boost by Alexander Klibanov at the MIT who showed emphatically (but not for the first time) that some enzymes (especially lipases) could function in organic solvents, thus broadening the substrate range to include water-insoluble substances[4].

For a while, in the early 1990s, the interest in the use of enzymes in organic synthesis increased at an almost exponential rate and two-volume works were needed even to summarize developments in the field[5]. Now, at the turn of the century, it is abundantly clear that the science of biotransformations has a significant role to play in the area of preparative chemistry; however, it is, by no stretch of the imagination, a panacea for the synthetic organic chemist. Nevertheless, biocatalysis is the method of choice for the preparation of some classes of optically active materials. In other cases the employment of man-made catalysts is preferred. In this review, a comparison will be made of the different methods available for the preparation of various classes of chiral compounds[6].

Obviously, in a relatively small work such as this it is not possible to be comprehensive. Preparations of bulk, achiral materials (e.g. simple oxiranes such as ethylene oxide) involving key catalytic processes will not be featured. Only a handful of representative examples of preparations of optically inactive compounds will be given, since the emphasis in the main body of this book, i.e. the experimental section, is on the preparation of chiral compounds. The focus on the preparation of compounds in single enantiomer form reflects the much increased importance of these compounds in the fine chemical industry (e.g. for pharmaceuticals, agrichemicals, fragrances, flavours and the suppliers of intermediates for these products).

The text of this short review article will be broken down into the following sections:

1. Hydrolysis of esters, amides, nitriles and oxiranes
2. Reduction reactions
3. Oxidative transformations
4. Carbon–carbon bond forming reactions.

In each of these areas the relative merits of biocatalysis versus other catalytic methodologies will be assessed. Note that the text is given an asterisk (*) when mention is made of a catalyst for a reduction or oxidation reaction that is featured in the later experimental section of this book.

1.1 HYDROLYSIS OF ESTERS, AMIDES, NITRILES AND OXIRANES

The enantioselective hydrolysis of racemic esters to give optically active acids and/or alcohols (Figure 1.1) is a well established protocol using esterases or lipases. In general, esterases from microorganisms or animal sources (such as

Pseudomonas putida esterase or pig liver esterase, (ple) or proteases (e.g. subtilisin) are employed in the reactions described in equation **(1)**, while lipases (e.g. *Candida antarctica* lipase) are more often used for transformations illustrated in

$$H_2O + R^*CO_2Me \xrightarrow{\text{Enz, } H_2O} R^*CO_2H + MeOH \tag{1}$$

$$H_2O + R^*OCOMe \xrightarrow{\text{Enz, } H_2O} R^*OH + MeCO_2H \tag{2}$$

$$H_2O + R^1CO_2R^{2*} \xrightarrow{\text{Enz, } H_2O} R^1CO_2H + R^{2*}OH \tag{3}$$

R* = chiral unit; Enz = esterase or lipase

Figure 1.1   Generalized scheme illustrating the hydrolysis of esters using enzymes.

equations **(2)** and **(3)**. Obviously in order to obtain optically active acid and/or alcohol the reaction is not taken to completion but stopped at about the halfway stage. The enantiomer ratio $E^{[7]}$ indicates the selectivity of the enzyme catalysed reaction. E values > 100 indicate highly enantioselective biotransformations. Typical resolutions are illustrated in Schemes $1^{[8]}$ and $2^{[9]}$. There have been

$$R^1CH(Me)CO_2Me \xrightarrow{i} R^1CH(Me)CO_2H + MeOH$$
$$(S)\text{-stereoisomer } E > 500$$

Scheme 1: Reagents and conditions: i) *Ps. putida* esterase H_2O.

$$F_5C_6-CH(OCOMe)CN \xrightarrow{i} F_5C_6-CH(OH)CN + MeCO_2H$$
$$(S)\text{-stereoisomer } E > 200$$

Scheme 2: Reagents and conditions: i) lipase LIP, H_2O, buffer pH 5–6.

models postulated for many of the popular enzymes (pig liver esterase, *Candida rugosa* lipase) in order better to **predict** the preferred substrate in a racemic mixture[10].

The ability of hydrolases to hydrolyse esters derived from primary alcohols in the presence of esters derived from secondary alcohols has been recognized (Scheme 3)[11].

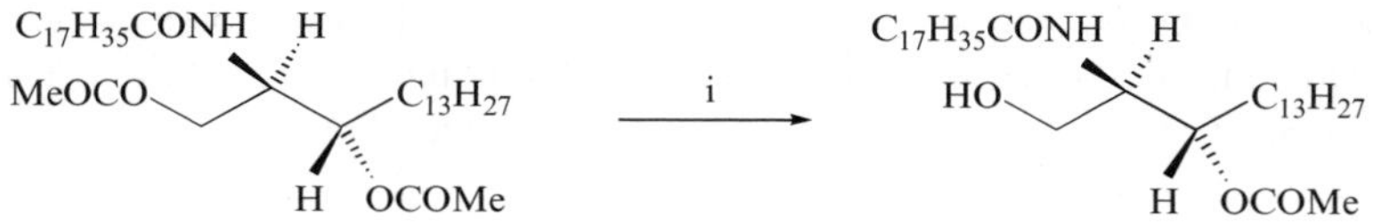

C$_{17}$H$_{35}$CONH H

MeOCO   C$_{13}$H$_{27}$

H OCOMe

$\xrightarrow{i}$

C$_{17}$H$_{35}$CONH H

HO   C$_{13}$H$_{27}$

H OCOMe

Scheme 3: <u>Reagents and conditions</u>: i) *Burkholderia cepacia* lipase, H$_2$O, buffer pH 7, decane.

However, the exquisite selectivity of hydrolase enzymes is, perhaps, best illustrated by their ability to produce optically active compounds from prochiral and *meso*-substrates. In both these cases a theoretical yield of 100 % for optically pure material is possible (Scheme 4)[12, 13].

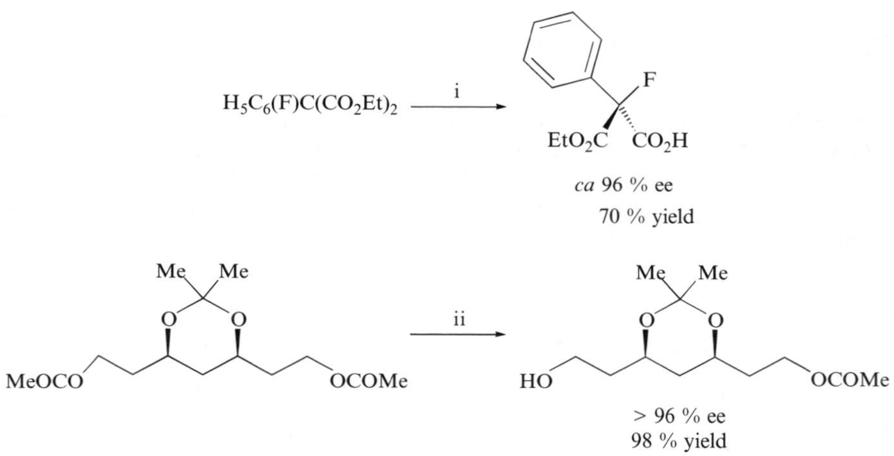

H$_5$C$_6$(F)C(CO$_2$Et)$_2$ $\xrightarrow{i}$

F

EtO$_2$C CO$_2$H

ca 96 % ee
70 % yield

Me Me

O O

MeOCO   OCOMe

$\xrightarrow{ii}$

Me Me

O O

HO   OCOMe

> 96 % ee
98 % yield

Scheme 4: <u>Reagents and conditions</u> i) Porcine pancreatic lipase, H$_2$O ii) *Ps. fluorescens* lipase, H$_2$O.

No other catalysts compete favourably with the enzymes in this type of work. Similarly lipases are the catalysts of choice for the enantioselective acylation of

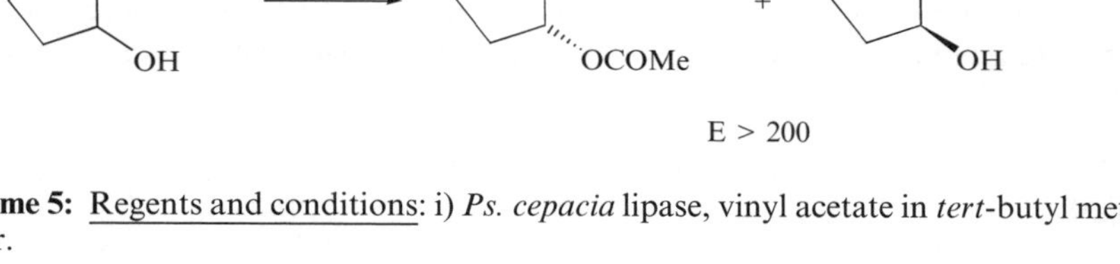

CO$_2$CH$_2$Ph
NH

OH

$\xrightarrow{i}$

CO$_2$CH$_2$Ph
NH

OCOMe

+

CO$_2$CH$_2$Ph
NH

OH

E > 200

Scheme 5: <u>Regents and conditions</u>: i) *Ps. cepacia* lipase, vinyl acetate in *tert*-butyl methyl ether.

a wide a variety of alcohols. This area of research has mushroomed since Klibanov's seminal studies clearly indicating that the procedure is exceedingly simple; a comprehensive review of the methodology is available[14]. A typical example of a resolution process involving enantioselective esterification using a lipase is shown in Scheme 5[15]. Furthermore, the mono-esterification of *meso*-diols represents an efficient way to generate optically active compounds (Scheme 6)[16].

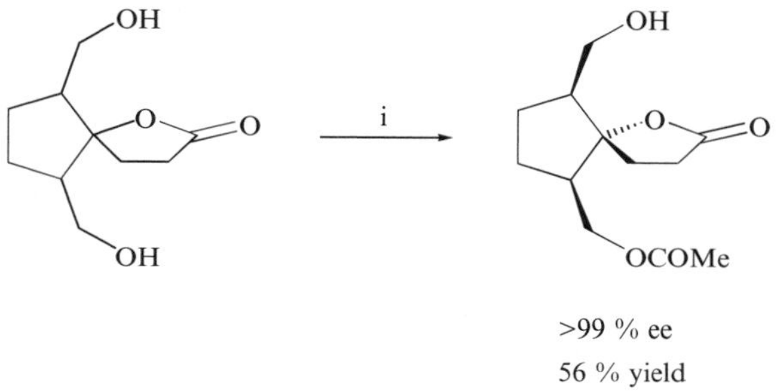

>99 % ee

56 % yield

Scheme 6: <u>Reagents and conditions</u>: i) *Ps. fluorescens* lipase, vinyl acetate in *n*-octane.

To a much smaller extent non-enzymic processes have also been used to catalyse the stereoselective acylation of alcohols. For example, a simple tripeptide has been used, in conjunction with acetic anhydride, to convert *trans*-2-acetylaminocyclohexanol into the (*R*),(*R*)-ester and recovered (*S*),(*S*)-alcohol[17]. In another, related, example a chiral amine, in the presence of molecular sieve and the appropriate acylating agent, has been used as a catalyst in the conversion of cyclohexane-1(*S*), 2(*R*)-diol into 2(*S*)-benzoyloxy-cyclohexan-1(*R*)-ol[18]. Such alternative methods have not been extensively ex-plored, though reports by Fu, Miller, Vedejs and co-workers on enantioselective esterifications, for example of 1-phenylethanol and other substrates using *iso*-propyl anhydride and a chiral phosphine catalyst will undoubtedly attract more attention to this area[19].

The chemo-, regio- and stereoselective hydrolysis of amides using enzymes (for example, acylases from hog kidney) has been recognized for many years. In the area of antibacterial chemotherapy, the use of an acylase from *Escherichia coli* to cleave the side-chain amide function of fermented penicillins to provide 6-aminopenicillanic acid *en route* to semi-synthetic penicillins has been taken to a very large scale (16 000 tonnes/year). The same strategy is used to prepare optically active amino acids. For instance, an acylase from the mould *Asper-gillus oryzae* is used to hydrolyse *N*-acyl DL-methionine to afford the L-amino acid and unreacted *N*-acyl-D-amino acid. The latter compound is separated, chemically racemized and recycled. L-Methionine is produced in this way to the extent of about 150 tonnes/year[1].

The hydrolysis of racemic non-natural amides has led to useful products and intermediates for the fine chemical industry. Thus hydrolysis of the racemic amide (2) with an acylase in *Rhodococcus erythrolpolis* furnished the (S)-acid (the anti-inflammatory agent Naproxen) in 42% yield and > 99% enantiomeric excess[20]. Obtaining the γ-lactam (−)-(3) has been the subject of much research and development effort, since the compound is a very versatile synthon for the production of carbocyclic nucleosides. An acylase from *Comamonas acidovorans* has been isolated, cloned and overexpressed. The acylase tolerates a 500 g/litre input of racemic lactam, hydrolyses only the (+)-enantiomer leaving the desired intermediate essentially optically pure (E > 400)[21].

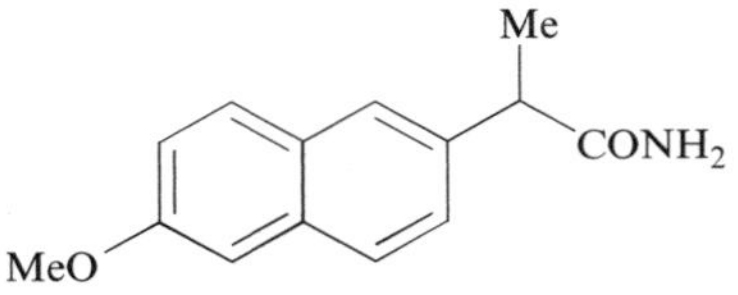

(2) 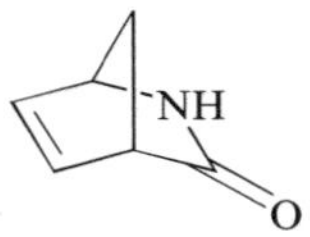   (−)-(3)

The enzyme-catalysed hydrolysis of epoxides has been reviewed[22]. Much of the early work featured liver microsomal epoxide hydrolases but the very nature and origin of these biocatalysts meant that they would always be limited to the small scale. In recent years the use of epoxide-hydrolase enzymes within organisms has become popular, with the fungus *Beauvaria sulfurescens* being featured regularly. For instance, incubation of styrene oxide with this organism provides (R)-1-phenylethanediol (45% yield; 83% ee) and recovered (R)-styrene oxide (34% yield; 98% ee)[23]. A particularly interesting example, shown in Scheme 7, is the stereoconvergent ring-opening of the racemic epoxide (4) which gives (R), (R)1-phenylpropane-1, 2-diol in 85% yield and 98% ee (one enantiomer of the epoxide suffers attack by water adjacent to the phenyl group, the other enantiomer is attacked by water at the carbon atom bearing the methyl group)[24].

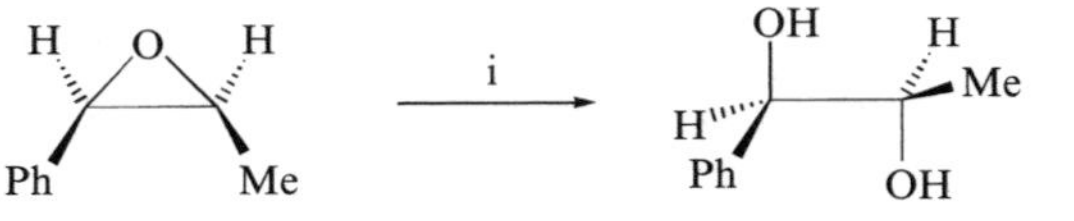

(±)-(4)

Scheme 7: Reagents and conditions i) *B. sulfurescens*, H_2O.

A major drawback in this area is that a portfolio of epoxide hydrolases is not available[25] and chemists remain reluctant to embark on processes which

involve the use of whole cells (such as *B. sulfurescens*). Not surprisingly, therefore, the use of a non-enzymic method for the kinetic resolution of terminal
epoxides and the stereoselective opening of *meso*-epoxides, involving salen–
cobalt complexes, has aroused interest. For example, use of the organometallic
catalyst in the presence of benzoic acid and cyclohexene epoxide afforded the
hydroxyester (**5**) (98 % yield; 77 % ee)[26].

OCOPh / OH

(**5**)

H / OCH$_2$Ph / HO$_2$C / CN

(**6**)

The same disadvantage (lack of commercially available enzymes, and the
consequent necessity for the employment of whole cells) dogs the otherwise
extremely useful biotransformation involving the hydrolysis of nitriles to the
corresponding amides (under the influence of a nitrile hydratase) or acids (by a
nitrilase). The conversion takes place under very mild conditions of temperature
and pH and some useful transformations have been recorded; for example the
cyanocarboxylic acid (**6**) (a precursor of the lactone moiety of mevinic acids) is
available from the corresponding prochiral dinitrile in good yield (60–70 %) and
high enantiomeric excess (88–99 % ee), on a multigram scale, over a period of 24
hours using *Rhodococcus* sp. SP361 or *Brevibacterium* sp. R312[27].

In summary, the formation of optically active compounds through hydrolysis reactions is dominated by biocatalysis mainly due to the availability and
ease of use of a wide variety of esterases, lipases and (to a lesser extent) acylases.
Epoxide ring-opening (and related reactions) is likely to be dominated by
salen–metal catalysts while enzyme-catalysed nitrile hydrolysis seems destined
to remain under-exploited until nitrilases or nitrile hydratases become commercially available.

1.2 REDUCTION REACTIONS

The balance between biocatalytic and other, organometallic-based, methodology is heavily biased in favour of the latter section when considering reduction reactions of importance in synthetic organic chemistry. Two areas will be
described to illustrate the point, namely the reduction of carbonyl groups and
the reduction of alkenes, not least since these points of focus complement
experimental work featured later in the book.

1.2.1   REDUCTION OF CARBONYL COMPOUNDS

It is well known that bakers' yeast is capable of reducing a wide range of ketones to optically active secondary alcohols. A recent example involves the preparation of the (R)-alcohol (7) (97% ee) (a key intermediate to (−)-norephedrine) from the corresponding ketone in 79% yield[28]. Other less well-known organisms are capable of performing similar tasks; for instance, reduction of 5-oxohexanoic acid with *Yamadazyma farinosa* furnishes (R)-5-hydroxyhexanoic acid in 98% yield and 97% ee[29].

(7)   (8)

However, the use of the whole cells of the microorganisms can lead to some difficulties. For instance, an aqueous solvent system is generally employed[30], certainly when the cells need to be in an active state and often when the cells are 'resting'[31]. The solubility of the substrate in the aqueous system can be problematic, as can the related transport of the starting material in to, and out from, the cytosol. At the end of the reaction, harvesting and disposal of the mycelial mass may be disconcerting, especially when considering large scale work. If a biocatalyst other than a readily available organism (such as bakers' yeast) is necessary then access to sterile equipment including fermenters is required, often considered a drawback for a person working in a conventional chemical laboratory. Thus, despite the various methods for improvement of particular protocols (including the immobilization of whole cell biocatalysts in alginate beads*)[32], whole-cell reduction reactions of carbonyl compounds remain, almost exclusively, in the small scale research arena.

It is possible to use isolated, partially purified enzymes (dehydrogenases) for the reduction of ketones to optically active secondary alcohols. However, a different set of complications arises. The new C–H bond is formed by delivery of the hydrogen atom from an enzyme cofactor, nicotinamide adenine dinucleotide (phosphate) NAD(P) in its reduced form. The cofactor is too expensive to be used in a stoichiometric quantity and must be recycled *in situ*. Recycling methods are relatively simple, using a sacrificial alcohol, or a second enzyme (formate dehydrogenase is popular) but the real and apparent complexity of the ensuing process (Scheme 8)[33] provides too much of a disincentive to investigation by non-experts.

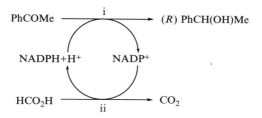

Scheme 8: <u>Reagents and conditions</u> i) dehydrogenase from *Lactobacillus* sp. ii) NADPH-dependent formate dehydrogenase.

Thus the methods of choice for the reduction of simple carbonyl compounds reside in the use of hydrogen and organometallic reagents*. Originally, reduction reactions using organorhodium complexes gained popularity. Thus hydrogenation of acetophenone in the presence of rhodium $(S),(S)$-2,4-bis(diphenylphosphinyl)pentane [(S,S)-BDPP or Skewphos] gave (S)-1-phenylethanol[34].

However, the employment of chiral ruthenium diphosphine–diamine mixed-ligand complexes has displaced much of the original experimentation to become the methodology of choice[35]. Such catalyst systems are prepared (sometimes *in situ*) by mixing a complex of BiNAP–RuCl$_2$ (**8**) with a chiral amine such as 1,2-diphenylethylenediamine (DPEN). In the presence of a base as co-catalyst such systems can achieve the reduction of a wide variety of alkyl arylketones under 1–10 atmospheres of hydrogen, affording the corresponding secondary alcohols in high enantiomeric excess[36]. A similar hydrogenation of tetralone using an iridium complex gave the (R)-alcohol (**9**) in 88% yield and 95% ee[37].

As an alternative to the use of hydrogen gas, asymmetric ruthenium-catalysed hydrogen transfer reactions have been explored with significant success*[38].

The reduction of dialkylketones and alkylaryl ketones is also conveniently accomplished using chiral oxazaborolidines, a methodology which emerged from relative obscurity in the late 1980s. The type of borane complex (based on (S)-diphenyl prolinol)[39] responsible for the reductions is depicted below (**10**). Reduction of acetophenone with this complex gives (R)-1-phenylethanol in 90–95% yield (95–99% ee)*[40]. Whilst previously used modified hydrides such as BiNAL–H (**11**), which were used in stoichiometric quantities, are generally unsatisfactory for the reduction of dialkylketones, oxazaborolidines

(9)   (10)   (11)

can be employed often with the production of secondary alcohols with high ee. For example *iso*-propylmethyl ketone and *tert*-butylmethyl ketone are good substrates giving secondary alcohols with >91% ee[41]. Alternatively oxaza-phosphinamides* and hydroxysulfoximines* have been used to control the stereochemistry of the reduction of simple ketones by borane.

Brook has effectively modified a procedure (introduced by Hosomi) which employs a trialkoxysilane as the stoichiometric reducing agent which, in the presence of amino acid anions reduces aryl alkyl ketones or diaryl ketones to the corresponding (*S*)-secondary alcohols, albeit in modest ee (generally 25–40%)*.

Much the same situation pertains to the asymmetric reduction of diketones and ketoesters. Thus, some years ago, a yeast reduction of the diketone (**12**) formed a key step in the preparation of important steroids (Scheme 9). Work in

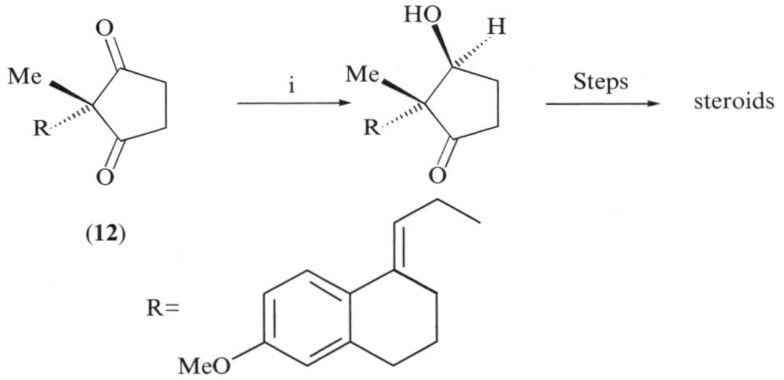

Scheme 9: Reagents and conditions: i) *Saccharomyces* sp., H_2O.

this area has continued, with closely related 2,2-disubstituted cyclopentane-1,3-diones[42] and other diketones. Indeed the diol (**13**) has been manufactured on a large scale by reduction of the corresponding hydroxydione using bakers' yeast[43]. The same microorganism is used in the reduction of a classical substrate, ethyl-3-oxobutanoate (aka ethyl acetoacetate), to give (*S*)-hydroxyester in a process optimised by Seebach*[44]. (Interestingly it has recently been shown that anaerobically grown bakers' yeast yields the corresponding (*R*)-alcohols with impressive optical purity [96–98% ee].)[45]

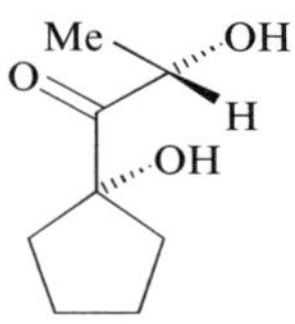

(**13**)

However, as for simpler carbonyl systems, organometallic catalysts offer a powerful alternative to biotransformations. By way of comparison methyl-3-oxobutanoate is reduced to the (R)-3-hydroxyester (> 99 % ee) quantitatively using (R)-BiNAP–RuCl$_2$ under 100 atmospheres of hydrogen[46]. A variation of this reaction using immobilized catalyst yields the chiral alcohol (92 % ee) at roughly the same rate[47], while Genêt's modification of the original procedure, preparing the catalyst *in situ* and employing a hydrogen pressure of one atmosphere, allows the reaction to be performed without special apparatus*. Note that other ligands have been employed for the ruthenium catalysed reduction of β-ketoesters. For example, a new diphosphine (BisP*), (+)-ephedrine and other amino alcohols (for asymmetric transfer hydrogenation of arylketones and β-ketoesters*) are described later, in the relevant experimental section.

Reduction of diketones such as pentane-2,4-dione using (R)-BiNAP–RuCl$_2$ under hydrogen (75–100 atm) gives the corresponding diol, in this case (R),(R)-2,4-pentanediol with an excellent diastereomer ratio (98 %) and optical purity (>99 %)[48].

When the dione has different terminal groups the Ru–BiNAP reduction can be selective towards one carbonyl group (Scheme 10)[49].

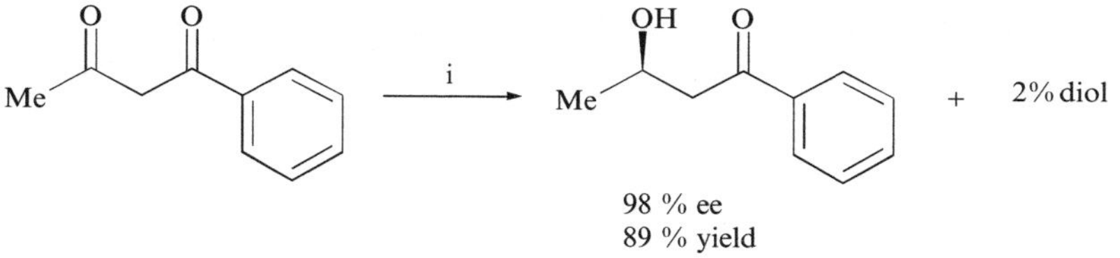

$$Me \quad \xrightarrow{\text{i}} \quad Me \quad + \quad 2\% \text{ diol}$$

98 % ee
89 % yield

Scheme 10: Reagents and conditions: i) H$_2$ (48 atm) [(R-BiNAP)RuCl(μ-Cl)$_3$] [NH$_2$ (C$_2$H$_5$)$_2$], MeOH, 50 °C.

1.2.2   REDUCTION OF ALKENES

Very few enzyme-catalysed reactions involving the reduction of alkenes have achieved any degree of recognition in synthetic organic chemistry. Indeed the only transformation of note involves the reduction of α, β-unsaturated aldehydes and ketones. For example, bakers' yeast reduction of (Z)-2-bromo-3-phenylprop-2-enal yields (S)-2-bromo-3-phenylpropanol in practically quantitative yield (99 % ee) when a resin is employed to control substrate concentration[50]. Similarly (Z)-3-bromo-4-phenylbut-3-en-2-one yields 2(S), 3(S)-3-bromo-4-phenylbutan-2-ol (80 % yield, >95 % ee)[51]. Carbon–carbon double bond reductases can be isolated; one such enzyme from bakers' yeast catalyses the reduction of enones of the type Ar − CH = C(CH$_3$) − COCH$_3$ to the corresponding (S)-ketones in almost quantitative yields and very high enantiomeric excesses[52].

One facet of the whole cell work that draws attention is the sometimes profitable operation of a cascade of reactions in the multi-enzyme portfolio of the microorganism. For instance (Scheme 11), the allylic alcohol (14) is reduced to the corresponding saturated compound in high yield and optical purity (though in a slow reaction) via the intermediacy of the corresponding enal and (S)-2-benzyloxymethylpropanal[53].

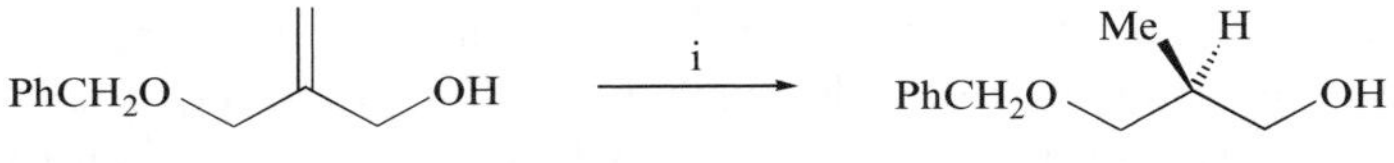

Scheme 11: <u>Reagents and conditions</u>: i) Bakers' yeast, 30 °C, 14 days.

Historically the biotransformations of cyclic enones have been important, not least Leuenberger's transformation of the appropriate cyclohexenedione into the saturated ketone (15), a precursor for tocopherol[54]. Similarly 2-methylcyclohex-2-enone is reduced by the microorganism *Yamadazyma farinosa* (also known as *Pichia farinosa*) to give a mixture of saturated alcohols and ketone; pyridinium chlorochromate oxidation of this mixture afforded 3(R)-methylcyclohexanone (95 % ee) in 67 % yield[55].

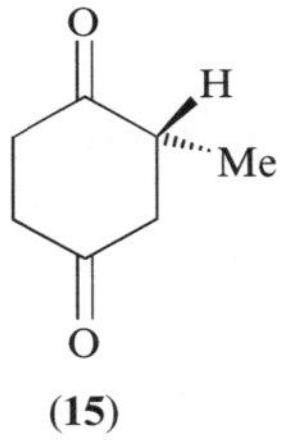

(15)

In the area of organometallic chemistry enantioselective hydrogenation of prochiral functionalised alkenes using chiral phosphine complexes of rhodium or ruthenium as catalysts has been extensively researched and, widely reported; the early work has been reviewed[56]. The first systems investigated involved organorhodium species particularly for the reduction of dehydroamino acid derivatives (Scheme 12)[57] but the emphasis shifted, some twenty years ago, to organoruthenium complexes, for example, the ruthenium–BiNAP system of Noyori[58]. The latter catalyst was found to be capable of catalysing the reduction of a wider range of substrates: for example, promoting the reduction of geraniol to (R)-citronellol (99 % ee) under hydrogen (100 atm) using methanol as the solvent and in the synthesis of benzomorphans and morphinans[59].

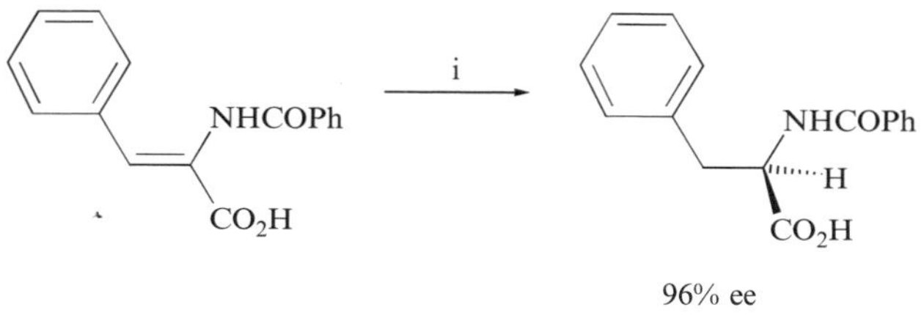

96% ee

Scheme 12: <u>Reagents and conditions</u>: i) [(BINAP)Rh(MeOH)$_2$]$^+$[ClO$_4$]$^-$ cat., 1 atm H$_2$, MeOH.

The broad range of alkenes undergoing asymmetric hydrogenation using ruthenium-based systems as catalysts has attracted the attention of chemists engaged in the synthesis of chiral biologically active natural products (Scheme 13)[60] and other pharmaceuticals (Scheme 14)[61]. α, β-Unsaturated phosphoric acids and esters have also proved to be suitable substrates for Ru(II)-catalysed asymmetric hydrogenation*[62].

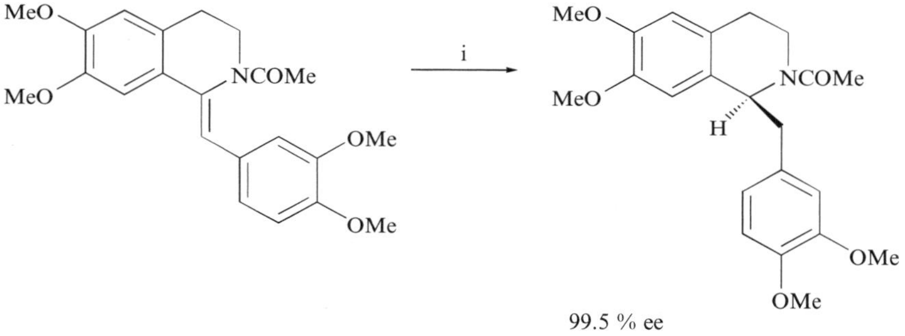

99.5 % ee

Scheme 13: <u>Reagents and conditions</u>: i) Ru–BiNAP (1 mol%), H$_2$ (4 atm) methylene chloride in MeOH.

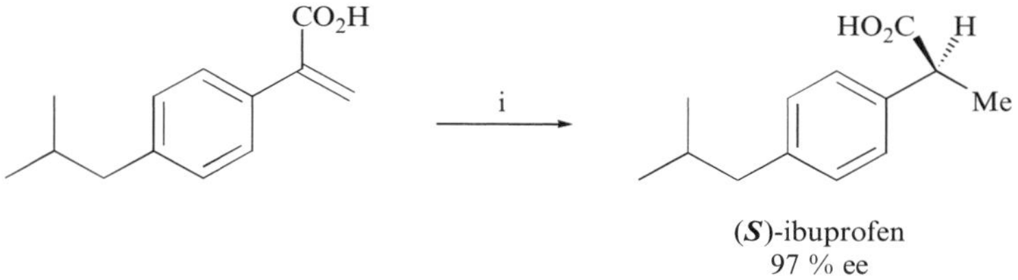

(S)-ibuprofen
97 % ee

Scheme 14: <u>Reagents and conditions</u>: i) Ru(S)–tetrahydroBiNAP (0.5 mol%), H$_2$ (100 atm), MeOH, 8 h.

Since these early days different ligands for rhodium complexes have been invented that more efficiently catalyse asymmetric reduction of a range of dehydroamino acids. One of the more popular ligands for general usage has been Burk's DuPHOS (Scheme 13)*[63].

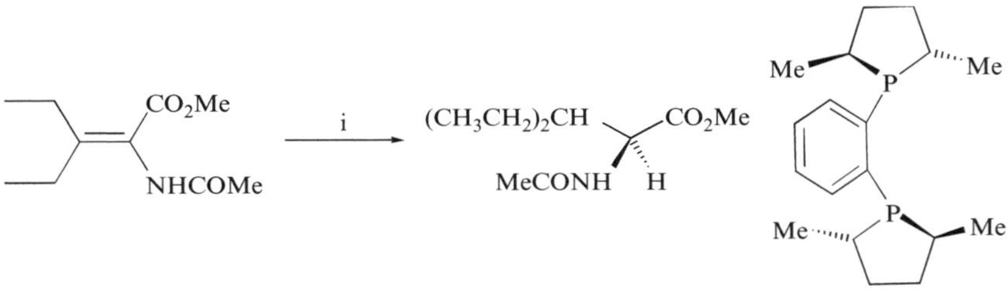

(S,S)–Me–DuPHOS

Scheme 15: <u>Reagents and conditions</u>: i) Rh–DuPHOS (0.2 mol%), H_2 (6 atm), benzene.

This strategy also gives access to a variety of non-natural α-amino acids. Furthermore, rhodium–DuPHOS complexes catalyse the asymmetric reduction of enol esters of the type $PhCH = CH - C(OCOCH_3) = CH_2$ to give (R)-2-acetoxy-4-phenylbut-3-ene (94 % ee)[64].

The use of chiral rhodium complexes fashioned from ferrocene derivatives has gained in popularity significantly in recent years*[65].

The portfolio of bisphosphine ligands for rhodium-catalysed asymmetric hydrogenation of dehydroamino acids is now becoming complemented by a set of bisphosphinite ligands*, typified by Chan's spirOP (16)[66] and carbohy-drate-based systems invented by Selke and RajanBabu[67]. The attraction of the use of the bisphosphinites lies in the simplicity of the preparation of the ligands (by reacting optically active diols with chlorophosphines in the presence of base)[68]. A remarkably selective one-pot procedure for sequential alkene and carbonyl reduction using chiral rhodium and ruthenium catalysts allows the preparation of amino alcohols with up to 95 % ee*[69].

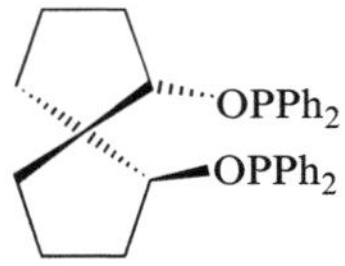

(16)

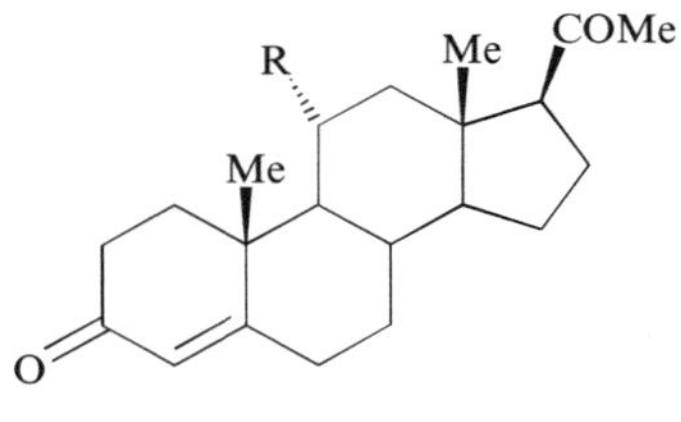

(17) R = H
(18) R = OH

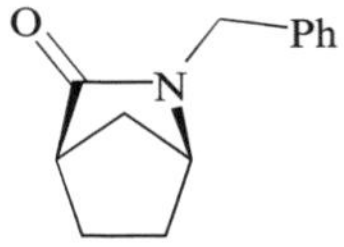

(19)

1.3   OXIDATIVE TRANSFORMATIONS

In this important area of synthetic chemistry honours are more equally shared between biocatalysis and other forms of catalysts, the latter being made up, almost invariably, of man-made organometallic species. Thus biotransformations are the preferred pathway for the hydroxylation of aliphatic, alicyclic, aromatic and heterocyclic compounds, particularly at positions remote from pre-existing functionality[70]. In contrast organometallic species are the catalysts of choice to convert alkenes into epoxides and diols. Both natural and non-natural catalysts are adept at the conversion of some sulfides into the corresponding sulfoxides and in performing stereoselective Baeyer–Villiger oxidations. Some of the details are provided hereunder.

The ability of microorganisms to convert alicyclic compounds into related alcohols by regio- and stereo-controlled hydroxylation at positions distant from regio- and stereo-directing functional groups was used extensively in the modification of steroids[711]. In a classical example the hydroxylation of progesterone (17) with *Rhizopus* sp. or *Aspergillus* sp. furnished the oxidized product (18), forming a key step in a highly efficient pathway to the anti-inflammatory steroids such as Betnovate[72]. Other complex alicyclic natural products and closely related compounds (e.g. taxanes)[73] have been selectively hydroxylated using some of the more easily handled organisms such as *Mucor* sp., *Absidia* sp. and *Cunninghamella* sp.

The selective monohydroxylation of heterocyclic compounds such as piperidine derivatives[74] and the γ-lactam (19)[75] have been studied. It is also been shown that hydroxylation of phenylcyclohexane can be effected using cytochrome P450 and the regioselectivity of hydroxylation can be altered by site-directed mutagenesis of the enzyme[76].

While undoubtedly powerful methodology, the major problem concerning enzyme-catalysed hydroxylation of alicyclic and saturated heterocyclic compounds is the unpredictability of the site of hydroxylation. Not surprisingly a start has been made to control the regioselectivity of microbial hydroxylation by using an easily-introduced and easily-removed directing group which, if such a suitable auxiliary could be found, would very conveniently promote hydroxylation at a set distance from the temporary appendage[77].

The hydroxylation of aromatic compounds using microorganisms is more predictable and a number of processes have been adapted to large scale, for example the preparation of 6-hydroxynicotinic acid[78] and (R)-2-(4-hydroxyphenoxy)propanoic acid[79], important intermediates to pesticides and herbicides respectively.

The biotransformation that has caught the imagination of many synthetic organic chemists involves the conversion of benzene and simple derivatives (toluene, chlorobenzene, etc.) into cyclohexadienediols (20) using a recombinant microorganism *E. coli* JM109. The one step oxidation, via reduction of the

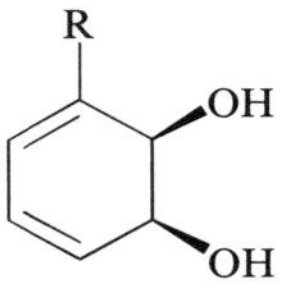

R = H, Me, Cl

(20)

corresponding dioxetane, is impossible to emulate using any other simple process. Cyclohexa-3,5-diene-1,2-diol produced in this way was used as a starting material in the polymer industry. The dienediol products from toluene and chlorobenzene are even more interesting being essentially optically pure and, for this reason, they have been used to prepare optically active morphinans, carbohydrate analogues, pancratistatin and *cis*-chrysanthemic acid, generally by selective transformations involving the two alkene bonds[80].

However, for the dihydroxylation of alkenes the microbiological method is not so effective and the biocatalytic methodology pales into insignificance compared with the powerful chemical technique introduced by Sharpless.

Also fifteen years of painstaking work and the gradual improvement of the system, the Sharpless team announced that asymmetric dihydroxylation (AD) of nearly every type of alkene can be accomplished using osmium tetraoxide, a co-oxidant such as potassium ferricyanide, the crucial chiral ligand based on a dihydroquinidine (DHQD) **(21)** or dihydroquinine (DHQ) **(22)** and methanesulfonamide to increase the rate of hydrolysis of intermediate osmate esters*[81].

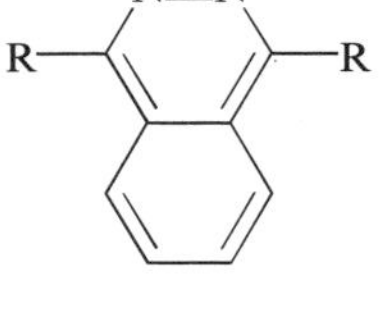

PHAL

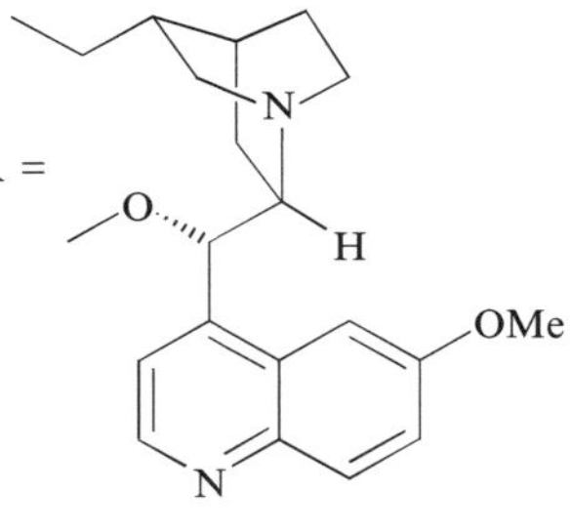

(21) DHQD

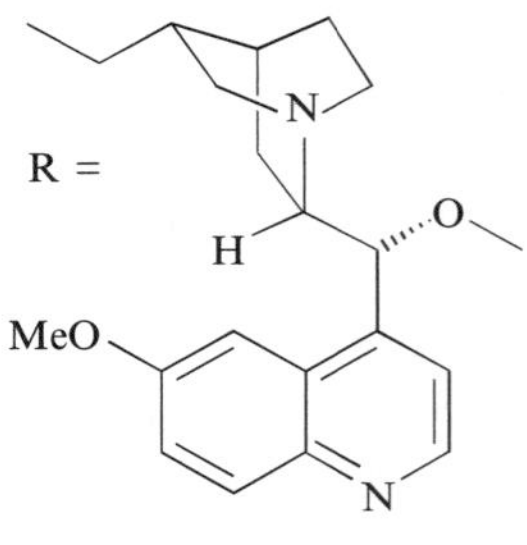

(22) DHQ

A wide range of alkenes undergo the Sharpless AD reaction and the stereochemistry of the product diols can be predicted with a high degree of certainty, in most cases, through a simple mnemonic device (Figure 1.2). Thus the DHQD derivatives supplied with the oxidant have become known as AD-mix β while the DHQ derivatives (with oxidant) comprise AD-mix α[81]. Chosen from the

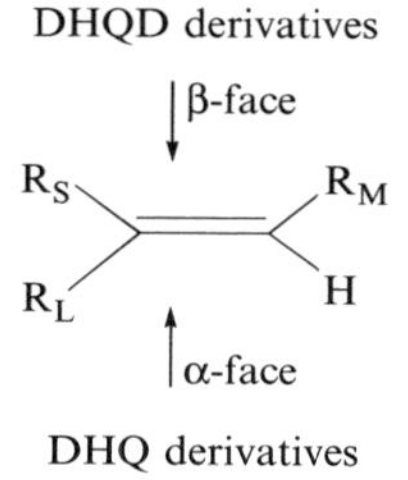

DHQD derivatives

DHQ derivatives

Figure 1.2   Predictive model for dihydroxylation of alkenes.

very many reports in the literature[82] three examples are given in Scheme 16[83]. A modification of the dihydroxylation reaction allows for the aminohydroxylation of alkenes and this reaction is also assuming an important role in organic synthesis*[84].

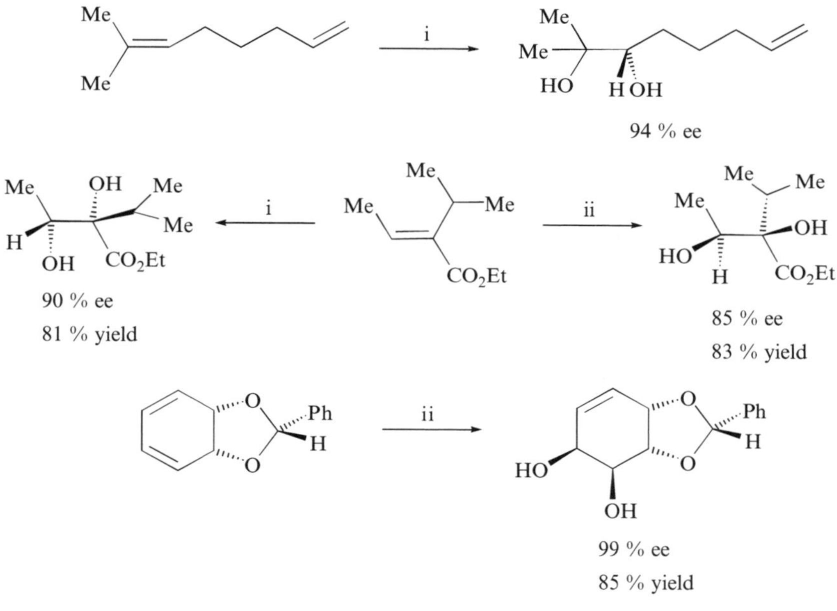

Scheme 16: Reagents and conditions: i) AD-mix α, $K_3Fe(CN)_6$, $MeSO_2NH_2$, t-BuOH, H_2O ii) AD-mix β, $K_3Fe(CN)_6$, $MeSO_2NH_2$, t-BuOH, H_2O.

The asymmetric dihydroxylation protocol was the second massive contribution by Professor Barry Sharpless to synthetic organic chemistry. The first procedure, introduced with Katsuki, involves the catalytic asymmetric epoxidation of allylic alcohols. A typical example is shown in Scheme 17, wherein (E)-allylic alcohol (**23**) is epoxidized with tert-butylhydroperoxide, in the presence of titanium tetra-isopropoxide and optically active diethyl tartrate to give the epoxyalcohol **(24)**[85]. (Note, however, that the isomeric (Z)-alkene undergoes asymmetric epoxidation much less efficiently.) Such reactions are rendered catalytic by the addition of 4Å molecular sieves to adsorb adventitious water which otherwise attacks the key component, the titanium tartrate complex. The sense of asymmetric epoxidation of E-allylic primary alcohols is highly predictable*. The preferred products of the Katsuki–Sharpless oxidation are shown in Figure 1.3. (Z)-Allylic alcohols undergo less predictable oxidation, as mentioned above.

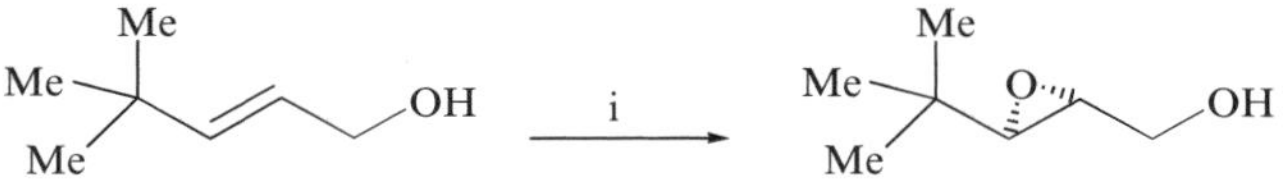

(23)   **(24)**

95 % ee

52 % yield

Scheme 17: Reagents and conditions: i) Ti(O–i-Pr)$_4$ (+)-diethyl tartrate, t-butylhydroperoxide, −20 °C.

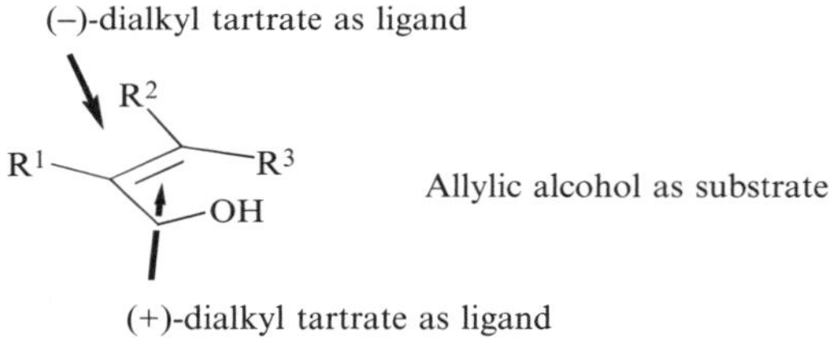

Figure 1.3   Oxidation of allylic alcohol using Ti(O − i-Pr)$_4$, TBHP and tartrate ligand.

Secondary allylic alcohols also undergo asymmetric epoxidation; in many cases, when the alcohol unit is attached to a stereogenic centre, kinetic resolution of the enantiomers takes place. This is particularly apparent for compounds of type **(25)**, where the two enantiomers are epoxidized at rates which are different by two orders of magnitude[86].

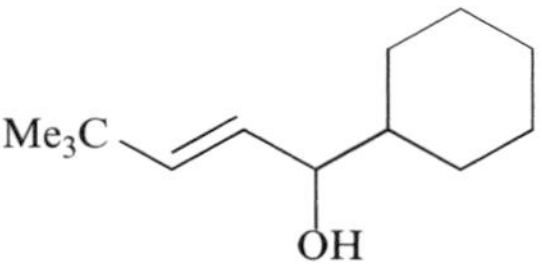

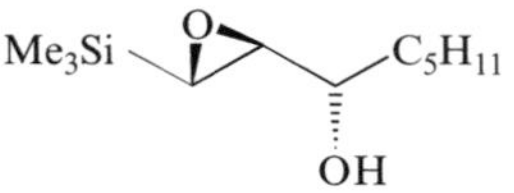

(25)   **(26)**

Similarly, racemic 1-trimethylsilyloct-(1E)-en-3-ol is epoxidized with Ti(O − i-Pr)$_4$, t-butylhydroperoxide and (+)-di-isopropyl tartrate at −20 °C to give the epoxide (26) (> 99 % ee, 42 %) and recovered (R)-unsaturated alcohol[87]. In general, when using (+)-tartrates, the (S)-enantiomer of the allylic alcohol will react faster.

The requirement for the presence of an adjacent alcohol group can be regarded as quite a severe limitation to the substrate range undergoing asymmetric epoxidation using the Katsuki–Sharpless method. To overcome this limitation new chiral metal complexes have been discovered which catalyse the epoxidation of nonfunctionalized alkenes. The work of Katsuki and Jacobsen in this area has been extremely important. Their development of chiral manganese (III)–salen complexes for asymmetric epoxidation of unfunctionalized olefins* has been reviewed[88].

A typical manganese–salen complex (27)[89] is capable of catalysing the asymmetric epoxidation of (Z)-alkenes (Scheme 18) using sodium hypochlorite (NaOCl) as the principle oxidant. Cyclic alkenes and α, β-unsaturated esters* are also excellent starting materials; for example indene may be transformed into the corresponding epoxide (28) with good enantiomeric excess[90]. The epoxidation of such alkenes can be improved by the addition of ammonium acetate to the catalyst system[91].

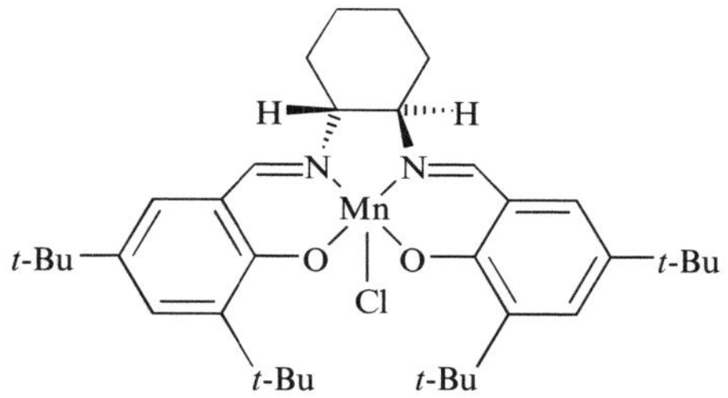

(27)

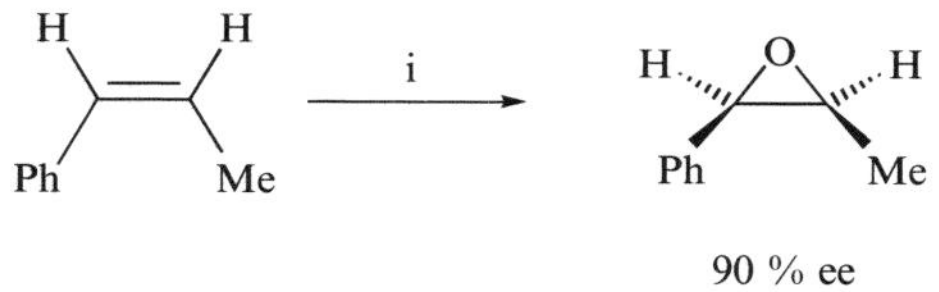

90 % ee

Scheme 18: Reagents and conditions: i) NaOCl, complex 27.

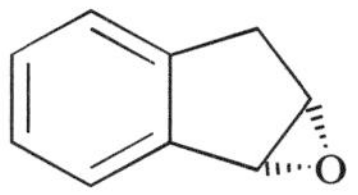

(28)

The asymmetric epoxidation of *E*-alkenes and terminal alkenes proved to be more difficult, though a recent finding, describing the use of a modified salen complex to epoxidize (*E*)-β-methylstyrene to form the corresponding epoxide in 83% ee, represents another important step forward. Alternatively, chiral (D_2-symmetric) porphyrins have been used, in conjunction with ruthenium* or iron, for efficient asymmetric oxidation of *trans*- and terminal alkenes[92].

The epoxidation of nonfunctionalized alkenes may also be effected by chiral dioxiranes*. These species, formed *in situ* using the appropriate ketone and potassium caroate (Oxone), can be formed from C-2 symmetric chiral ketones (**29**)[93], functionalized carbohydrates (**30**)[94] or alkaloid derivatives (**31**)[95]. One example from the laboratories of Shi and co-workers is given in Scheme 19.

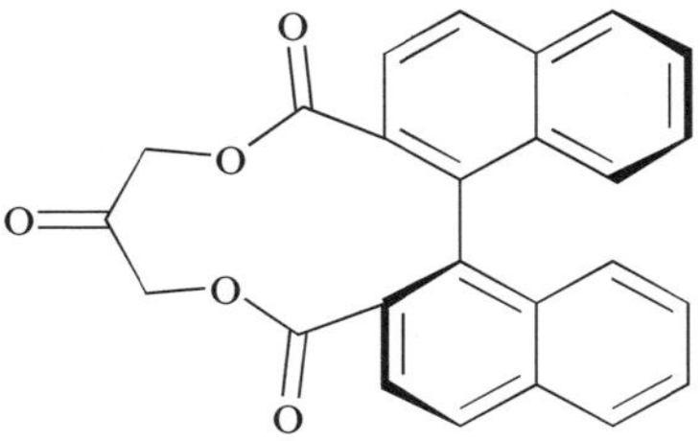

(**29**) 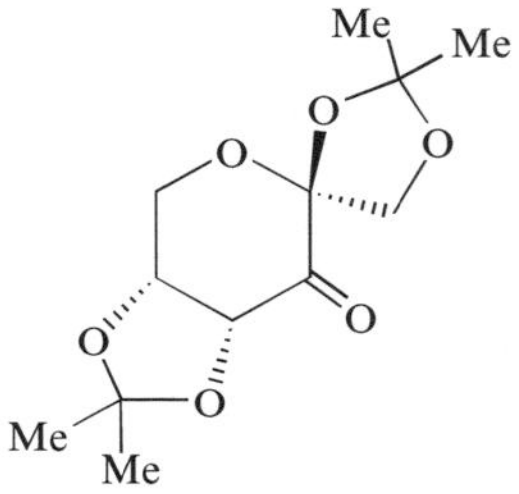   (**30**)

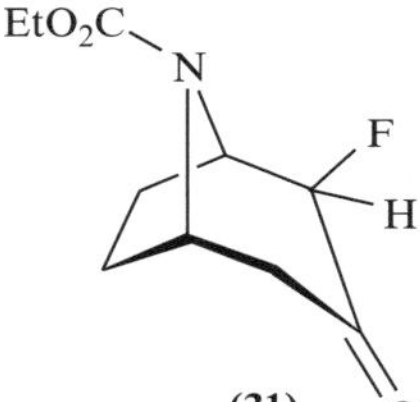

(**31**)

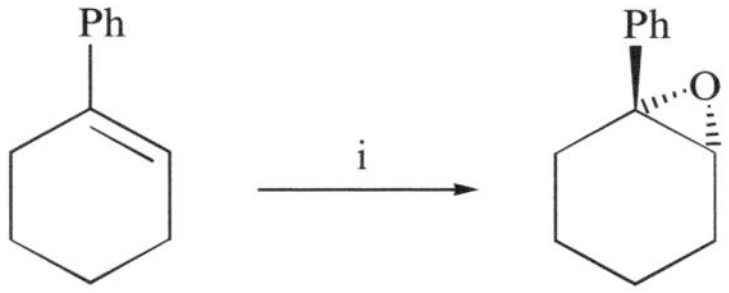

Scheme 19: Reagents and conditions: i) Oxone, NaHCO$_3$, CH$_3$CN, ketone (**30**) (30 mole%), −10 °C.

Historically, the asymmetric synthesis of epoxides derived from electron-poor alkenes, for example α, β-unsaturated ketones, has not received as much attention as the equivalent reaction for electron-rich alkenes (*vide supra*). However, a recent flurry of research activity in this area has uncovered several new methods. For example, Enders has shown that oxygen in the presence of diethylzinc and *N*-methyl ephedrine converts enones into epoxides in excellent yields and very good enantiomeric excesses* (up to 92%)[96]. Alternatively, Jackson *et al.* have reported the employment of *tert*-butyl hydroperoxide as the oxidant together with catalytic amounts of dibutyl magnesium and diethyl tartrate. Chalcones are oxidized to the corresponding epoxides under these conditions in yields varying between 40–60% and good to excellent enantiomeric excess[97].

For a similar series of chalcone derivatives the use of aqueous sodium hypochlorite in a two phase system (with toluene as the organic solvent) and the quinine derivative (**32**) as a chiral phase-transfer catalyst, produces epoxides with very good enantiomeric excesses and yields[98].

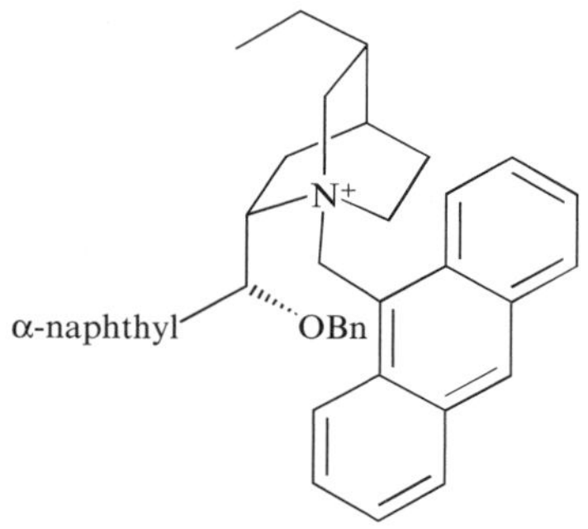

(**32**) 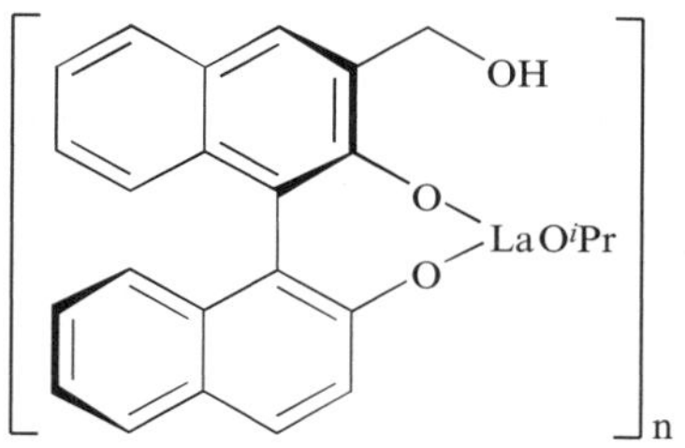   (**33**)

However, the two methods of choice for the oxidations of α, β-unsaturated ketones are based on lanthanoid–BINOL complexes* or a biomimetic process based on the use of polyamino acids as catalysts for the oxidation*[99].

In the first of these techniques the lanthanoid complex (**33**) (5–8 mol%) is used as the organometallic activator in cumene hydroperoxide or *tert*-butyl hydrogen peroxide-mediated oxidation of chalcone (epoxide yield 99%; 99% ee)* or the ketone (**34**) (Scheme 20)[100].

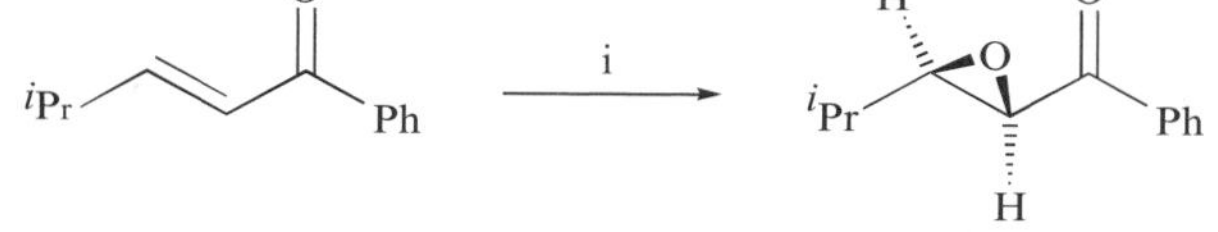

(**34**)

94 % ee
95 % yield

Scheme 20: Reagents and conditions: i) cumene hydrogen peroxide, 4Å molecular sieve, THF, complex (**33**).

The biomimetic protocol was invented by Juliá and Colonna, and involves the use of polyamino acids (such as poly-(L)-leucine) as the catalysts for peroxide oxidation of chalcones, styryl alkyl ketones and conjugated alkenones. The substrate range is broad, especially when using immobilized catalysts and an organic solvent containing the substrate, urea–hydrogen peroxide and an organic base (Scheme 22)[101].

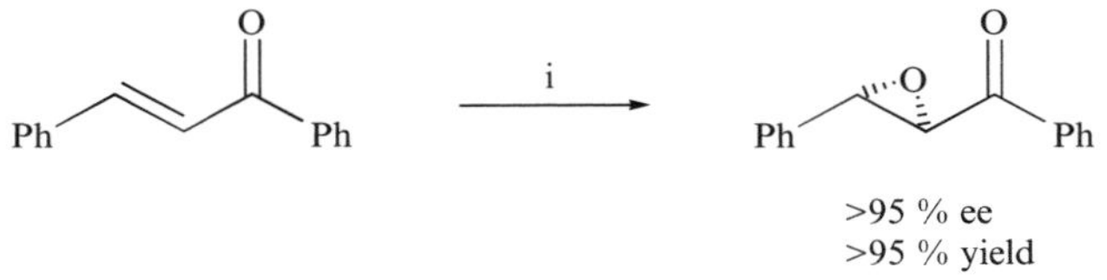

>95 % ee
>95 % yield

Scheme 22: Reagents and conditions: i) poly-L-leucine, urea–hydrogen peroxide, THF, diazabicycloundecene.

Neither biocatalysts nor non-natural catalysts have yet been found that provide a robust method of choice for stereocontrolled conversion of a ketone to an ester (or lactone) via a Baeyer–Villiger reaction. Thus whole cell biocatalysts can be used for very elegant transformations (Scheme 23)[102] but the microorganisms (such as *Acinetobacter* sp.) need to be grown and harvested before use (an anathema to most organic chemists, particularly for organisms which are potentially pathogenic) and the use of the relevant isolated enzymes (Baeyer–Villiger monooxygenases) is plagued with the problem of cofactor (NADH or NADPH) recycling and, often, progressive poisoning of the catalytic action of the enzyme by the product as it is formed[103]. The cloning of useful Baeyer–Villiger monooxygenases into bakers' yeast may give, in time, more widely-available, easily-used microorganisms[104].

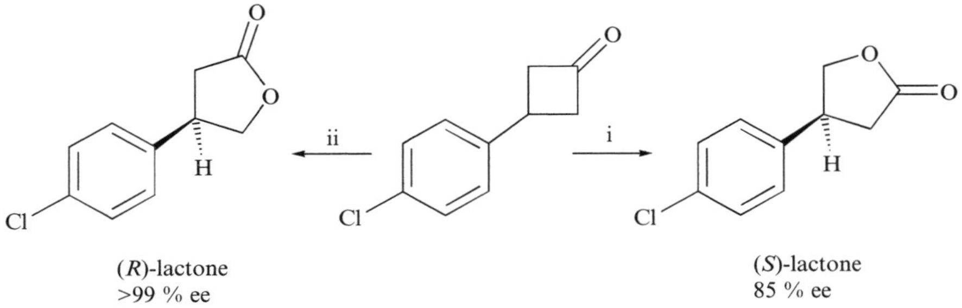

(R)-lactone   (S)-lactone
>99 % ee   85 % ee

Scheme 23: Reactions and conditions: i) *Acinetobacter calcoaceticus*, H_2O ii) *Cunninghamella echinulata*, H_2O.

3-Phenylcyclobutanone has been a substrate for a copper-catalysed Baeyer–Villiger oxidation. Thus the complex (**35**), (*ca* 1 mole%) in conjunction with

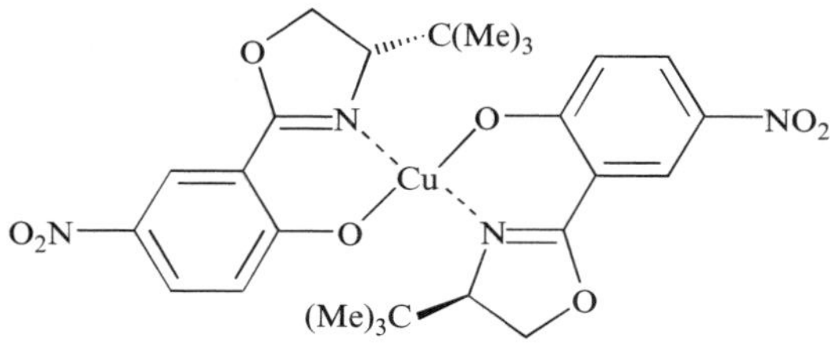

(35)

pivaldehyde in benzene under an atmosphere of oxygen gives a high yield of the (S)-γ-lactone but in only 44 % ee[105]. Similarly stereoselective oxidation of 3-hydroxymethylcyclobutanone has been accomplished with dialkyl tartrate/titanium complexes and *tert*-butyl hydroperoxide (conditions similar to those used in Sharpless asymmetric epoxidations). However, yields are modest and the enantiomeric excess of the (R)-lactone was just 75 %[106].

In contrast to the situation with the Baeyer–Villiger oxidation, synthetic chemists have a choice of both enzymatic or non-enzymatic methods for the oxidation of sulfides to optically active sulfoxides with good to excellent yields and enantiomeric excesses.

Thus a number of enzymes have been shown to be able to control the oxidation of sulfides to optically active sulfoxides; most extensive investigations have concentrated on mono-oxygenases (e.g. from *Acinetobacter* sp., *Pseudomonas putida*) and haloperoxidases[107] (from *Caldariomyces fumago* and *Corallina officinalis*). A comparison of the methodologies[108] led to the conclusion that the haloperoxidase method was more convenient since the catalysts are more readily available (from enzyme suppliers), the oxidant (H_2O_2) is cheap and no cofactor recycling is necessary with the haloperoxidases. Typical examples of haloperoxidase-catalysed reactions are described in Scheme 24.

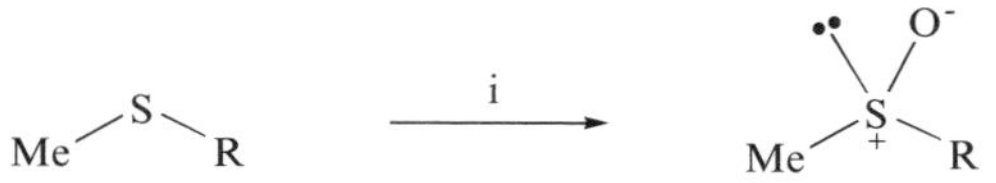

R = isopropyl, allyl,   >98 % ee
pentyl, cyclopentyl   75–98 % conversion

Scheme 24: Reactions and conditions: i) Chloroperoxidase from *Caldariomyces fumago*, H_2O_2, halide ion, H_2O.

Of several procedures for the stereoselective oxidation of sulfides using organometallic complexes*, two adaptations of Kagan's original process have gained prominence. In the first method the diol (36) is reacted with $Ti(O^iPr)_4$ to form the catalyst. With cumyl hydroperoxide as the stoichiometric oxidant, methyl *para*-tolyl sulfide was converted into the optically active sulfoxide in 42 % yield (98 % ee)[109].

In the second noteworthy adaptation of the Kagan method, Reetz and co-workers utilized the dinitrooctahydronaphthol (37). Oxidation of methyl *para*-tolylsulfide under similar conditions to those in the above paragraph furnished the optically active sulfoxide (86% ee)[110].

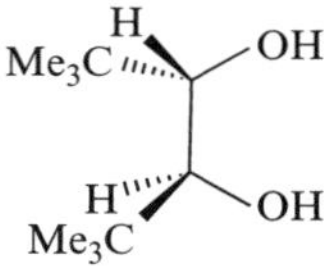

(36) 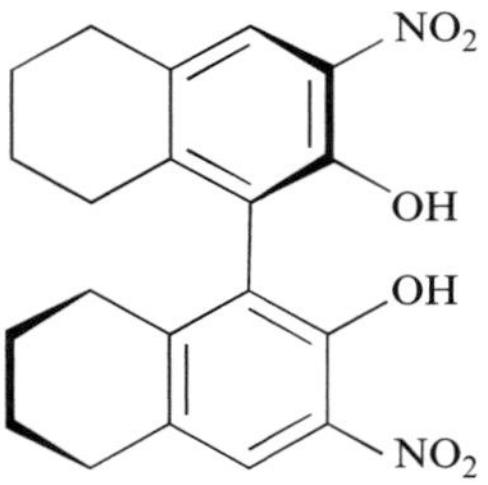   (37) 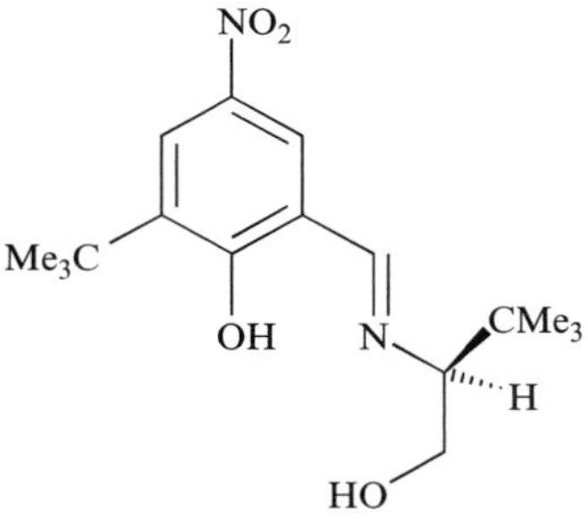   (38)

In addition, a recent report details a very efficient nonenzymatic method for the asymmetric oxidation of sulfides; this employs an organo-vanadium species featuring the imine (38) (Scheme 25)[111]. A second, complementary strategy for the preparation of optically active sulfoxides involves the enantioselective oxidation of racemic sulfoxides.*

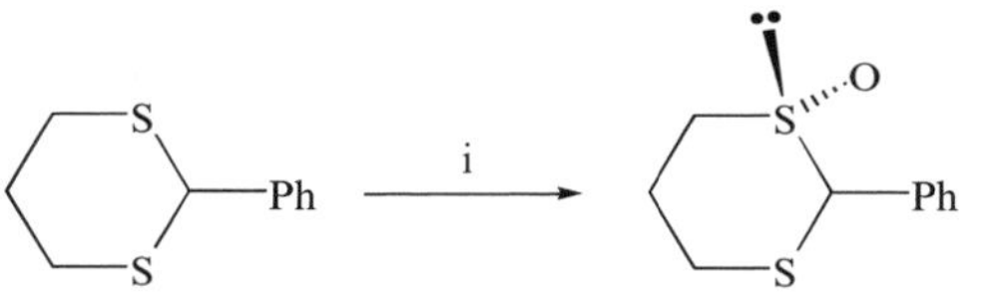

85 % ee

Scheme 25: Reagents and conditions: i) VO(acac)$_2$, compound (38), H$_2$O$_2$, H$_2$O, CH$_2$Cl$_2$.

1.4   CARBON–CARBON BOND-FORMING REACTIONS

In the arena of carbon–carbon bond-forming reactions, obviously a central feature in synthetic organic chemistry, the number of nonbiocatalytic methods in regular use far outweighs the small portfolio of biotransformations that can be considered to be available for general employment.

Indeed the only conversion where biocatalysis should be seriously considered is the transformation of aldehydes into optically active cyanohydrins[112]. For example, the conversion of aryl aldehydes into the appropriate (*R*)-cyanohydrins using almond meal may be accomplished in quantitative yield and gives products of high optical purity[113]. The method is much less successful for the vast majority of ketones.

(S)-Cyanohydrins are formed from a wide range of alkyl and aryl aldehydes (and also some methyl ketones) often in good yield and high enantiomeric excess using the enzyme (hydroxynitrile lyase) from *Hevea brasiliensis*[114]. The same range of substrates and the same cyanohydrins ((S)-configuration) are formed on catalysis of the addition of HCN using the hydroxynitrile lyase from *Manihot esculenta*. This enzyme has been cloned and over-expressed in *E. coli*[115].

A biomimetic method using a cyclic dipeptide (**39**) is available. In the presence of HCN in toluene containing 2 mole% of (**39**), benzaldehyde is converted into the (R)-cyanohydrin in 97% yield (97% ee)[116].

(**39**)   (**40**)

Complexation of an amino acid derivative with a transition metal to provide a cyanation catalyst has been the subject of investigation for some years. It has been shown that the complex formed on reaction of titanium(IV) ethoxide with the imine (**40**) produces a catalyst which adds the elements of HCN to a variety of aldehydes to furnish the (R)-cyanohydrins with high enantioselectivity[117]. Other imines of this general type provide the enantiomeric cyanohydrins from the same range of substrates[117].

The addition of trimethylsilyl (TMS) cyanide to aldehydes produces TMS-protected cyanohydrins. In a recent investigation a titanium salen-type catalyst has been employed to catalyse trimethylsilylcyanide addition to benzaldehyde at ambient temperature[118]. Several other protocols have been published which also lead to optically active products. One of the more successful has been described by Abiko *et al.* employing a yttrium complex derived from the chiral 1,3-diketone (**41**)[119] as the catalyst, while Shibasaki has used BINOL, modified so as to incorporate Lewis base units adjacent to the phenol moieties, as the chiral complexing agent[120].

The aldol reaction is of fundamental importance in organic chemistry and has been used as a key reaction in the synthesis of many complex natural products. There are biocatalysts for this reaction (aldolases) and one (rabbit muscle

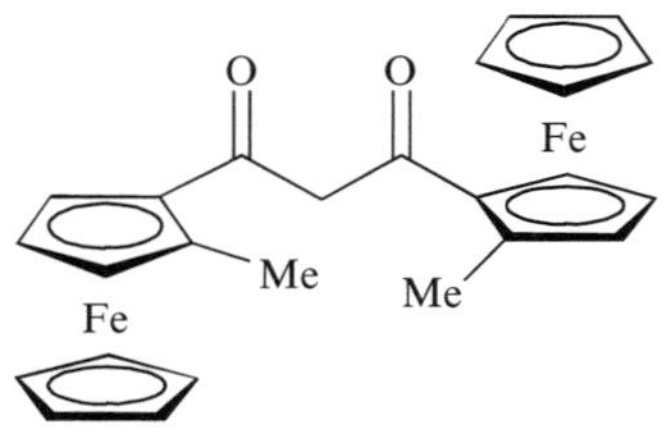

(41)

aldolase, RAMA) has been quite widely used for the preparation of carbohydrates and closely related compounds. For example, the azidotetraol (42) (a precursor of novel cyclic imine sugars active as α-fucosidase inhibitors) has been prepared by coupling dihydroxyacetone monophosphate and 2-azido-3-hydroxypropanal using RAMA as the catalyst, followed by dephosphorylation (Scheme 40)[121].

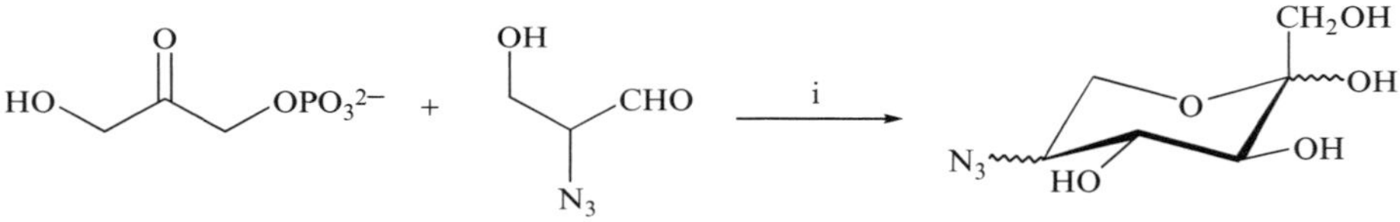

(42)

Scheme 40: Reagents and conditions: i) RAMA, H_2O then dephosphorylation using acid phosphatase.

Other aldolases, from microorganisms, have been cloned and overexpressed. For instance, L-threonine aldolase from *Escherichia coli* and D-threonine aldolase from *Xanthomonus orysae* have been obtained and used to prepare β-hydroxy-α-amino acid derivatives[122].

On moving away from carbohydrate chemistry one finds that non-natural catalysts are the materials of choice for the promotion of the classical aldol reaction and more recently-discovered variants. A wide range of methods are available and a small selection of these is described below.

One of the most widely studied aldol-type reactions is the Mukaiyama coupling of enol silanes of various types to aldehydes, catalysed by Lewis acids (notably organotin, organoboron, organotitanium and organocopper species). A typical example of the stereocontrolled coupling of an aromatic or aliphatic aldehyde and a silylthioketene acetal is described in Scheme 41. The products are generally obtained in 70–80% yield with a good to excellent diastereomeric excess of the *syn* isomer in 90–100% ee on using 10–30 mol% of the catalyst (43)[123].

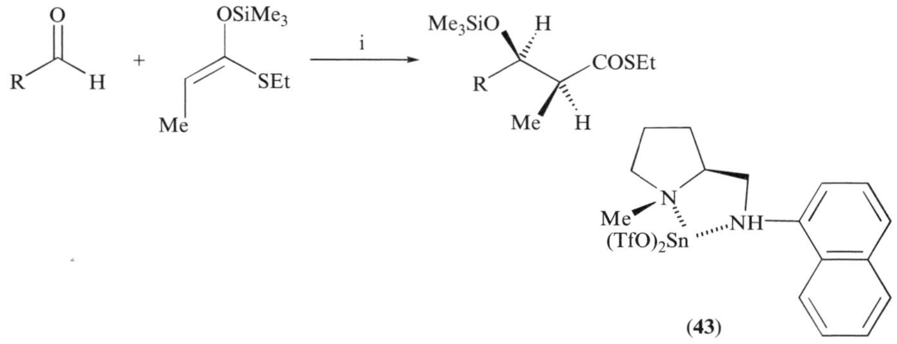

(43)

Scheme 41: Reagents and conditions: i) C_2H_5CN, $-78\,^\circ C$, 10–30 mol% complex (43).

Of the catalysts that are based on boron, the Masamune oxazaborolidines (44) are typical, being able to promote aldol reactions of the type described in Scheme 42[124].

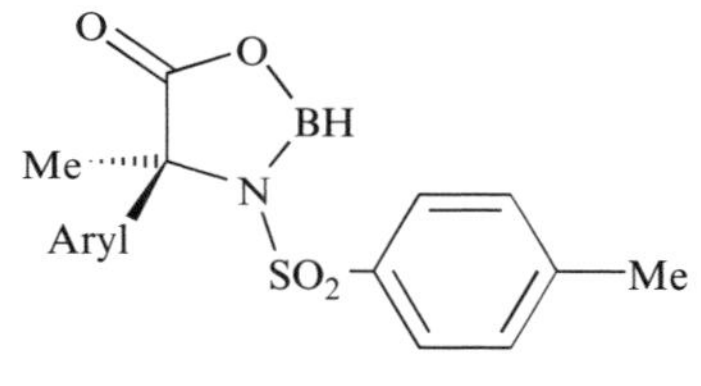

(44)

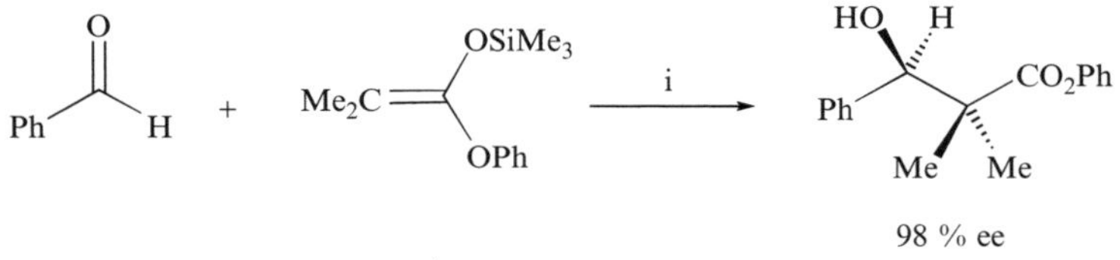

98 % ee

Scheme 42: Reagents and conditions: i) 20 mol% catalyst (44).

From the organotitanium family of catalysts BINOL–TiCl$_2$ and BINOL–Ti(^{i}PrO)$_2$ catalysts have been complemented by related catalysts of type (45) introduced by Carreira[125]. The simple enol silane (46) adds to a variety of aldehydes in high yields and excellent enantiomeric excesses, using as little as 0.5 mol% of the catalyst (45). The reacting aldehydes can bear some other functional groups, such as *tert*-butyldimethyl silyl ether moieties.

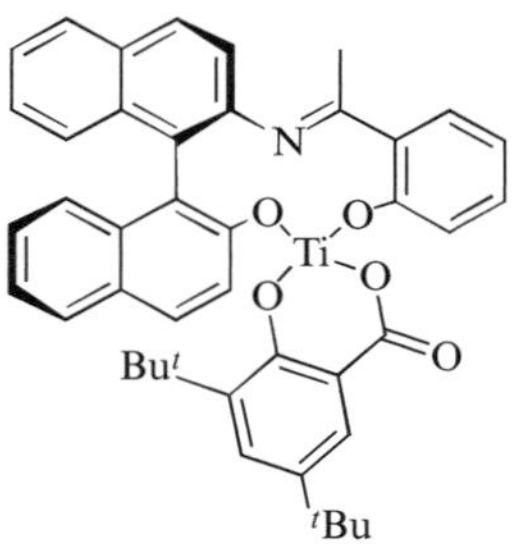

(45)

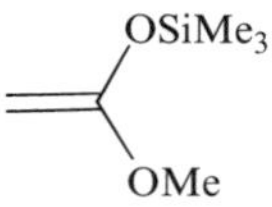

(46)

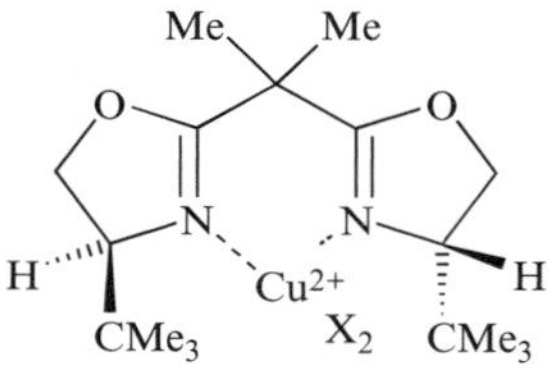

(47): X = CF$_3$SO$_3^-$

(48): X = [SbF$_6$]$^-$

The utilization of copper complexes (47) based on bisisoxazolines allows various silyl enol ethers to be added to aldehydes and ketones which possess an adjacent heteroatom: e.g. pyruvate esters. An example is shown is Scheme 43[126]. C$_2$-Symmetric Cu(II) complexes have also been used as chiral Lewis acids for the catalysis of enantioselective Michael additions of silylketene acetals to alkylidene malonates[127].

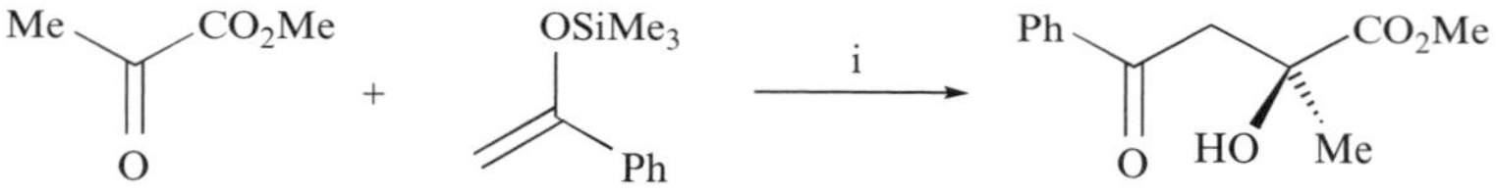

Scheme 43: Reagents and conditions: i) CH$_2$Cl$_2$, $-78\,°$C, 10 mol% catalyst (47).

Direct asymmetric aldol reactions, that is between aldehydes and unmodified ketones has been accomplished using a lanthanum trilithium tri(binaphthoxide) complex[128].

One of the key features of such stereocontrolled aldol reactions is the predictability of the absolute stereochemistry of the enantiomers (or diastereomers) that will be formed as the major products. The preferred intermediate for an archetypal aldol reaction, proceeding by way of a metal enolate, can be tracked using the Zimmerman–Traxler transition state and the results from the different variations of the aldol reaction can be interpreted from similar reasoning, and hence predictions made for analogous reactions[129].

The second well-known and much-used carbon–carbon bond forming reaction is a [4 + 2]-cycloaddition, the Diels–Alder reaction. Very many chiral Lewis acid catalysts have been used to promote this reaction and a *pot-pourri* of organo-aluminium, -boron and -copper catalysts are described, in brief, below.

The first organoaluminium complex that catalysed a Diels–Alder reaction was formed from menthol and ethylaluminium dichloride. This finding was complemented by work of Corey who showed that the aluminium–diamine complex (49) was effective for controlling the stereochemistry of Diels–Alder reactions involving cyclopentadiene and acryloyl and crotonyl amides (e.g.

$CH_3CH = CHCONR_2$). Later investigations showed the catalysts were also effective stereocontrolling systems for the coupling of a maleimide to an acyclic diene (Scheme 44)[130].

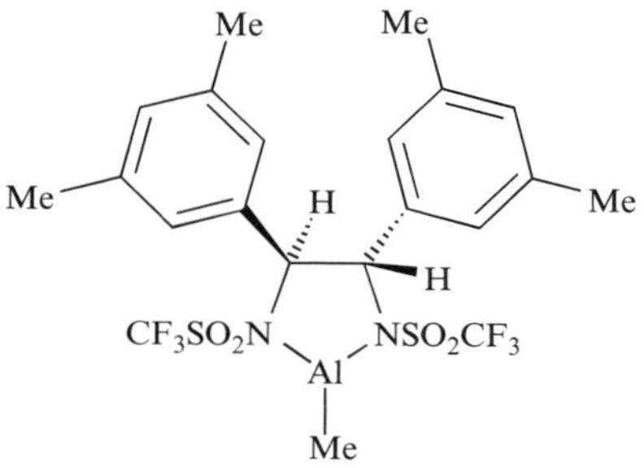

(49) 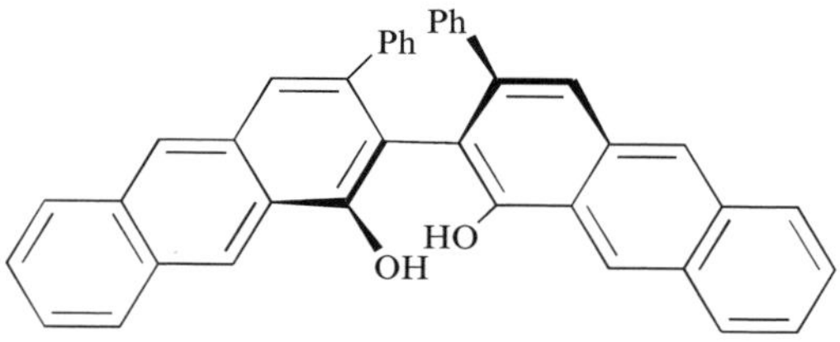   (50)

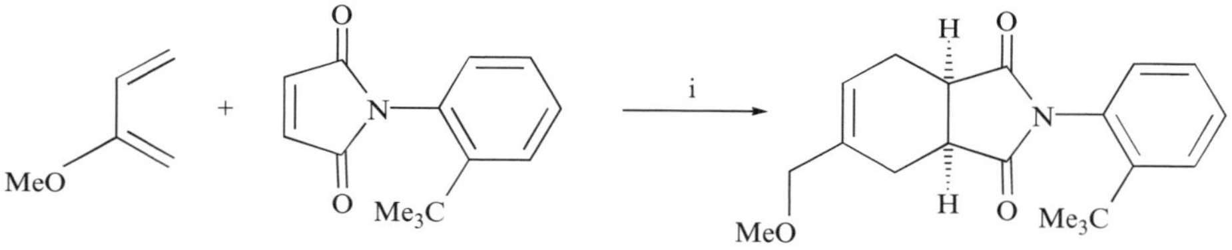

95 % ee

Scheme 44: Reagents and conditions: i) CH_2Cl_2, $-78\,^{\circ}C$, 10–20 mol% catalyst (49).

The biaryl compound (50) forms a complex with diethylaluminium chloride to provide a catalyst able to promote enantioselective reaction between cyclopentadiene and methacrolein or acrylates (Scheme 45). The addition of di-*tert* butyl 2,2-dimethylmalonate to the reaction mixture was found to enhance the enantiomeric excess of the product[131].

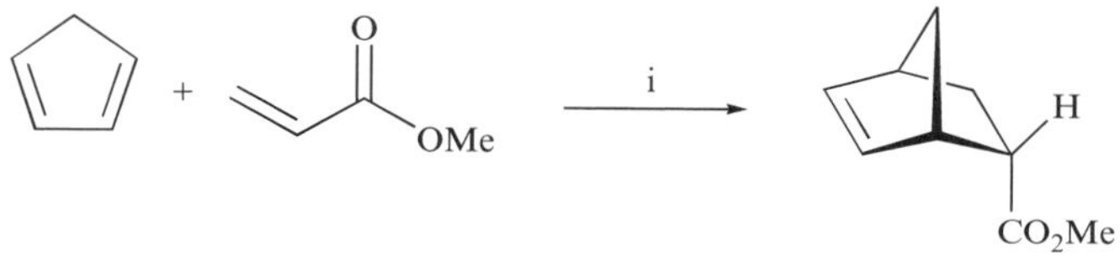

>99 % ee
76 % yield

Scheme 45: Reagents and conditions: i) $(Me_3CO_2C)_2CMe_2$ 50 mol%, Et_2AlCl and compound (51) 10 mol% each, CH_2Cl_2, $-78\,^{\circ}C$ to $-40\,^{\circ}C$.

Acyloxyboron complexes and oxazaborolidines have been shown to catalyse Diels–Alder reactions featuring aldehydes as one component: for example, the complex (51) allows the coupling of cyclopentadiene and α-bromoacrolein in high yield to give a product of high optical purity (Scheme 46)[132]. The immobilized catalyst system of this genre, recently introduced by Itsuno, is

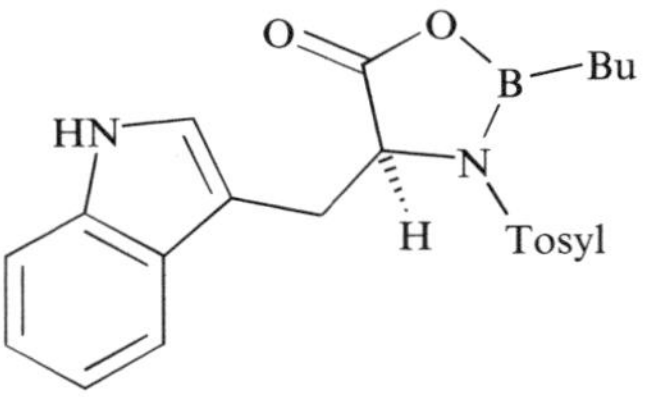

(51) 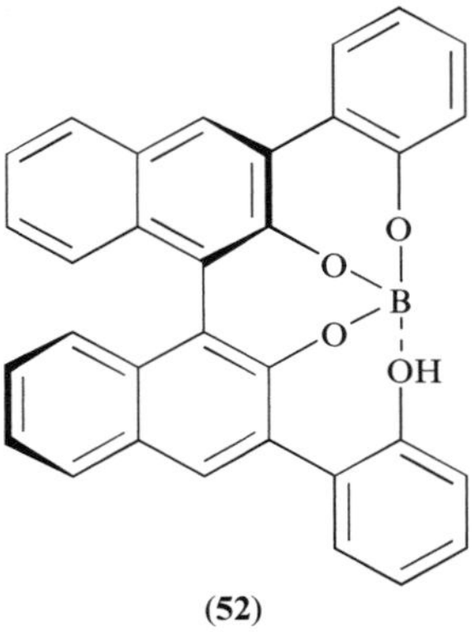   (52)

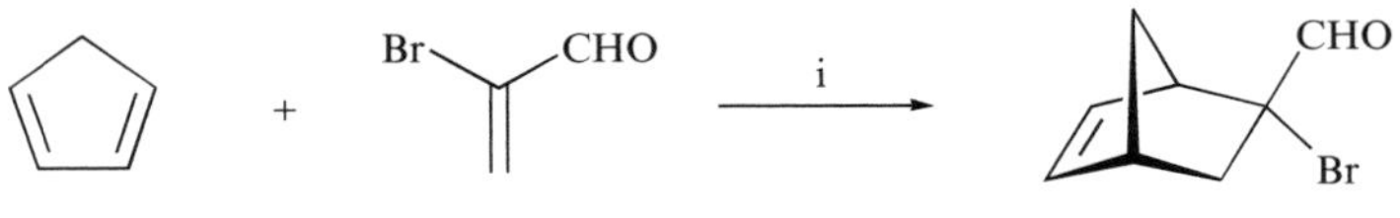

99 % ee
95 % yield

Scheme 46: <u>Reagents and conditions</u>: i) CH_2Cl_2, $-78\,°C$, 5–10 mol% catalyst (51).

also worthy of note[133]. In a further development Brønsted acid-assisted chiral Lewis acids such as compound (52) were shown to promote stereocontrolled reactions of dienes with a range of α, β-unsaturated aldehydes[134].

Copper(II)-bis(oxazoline) complexes (48) are robust, valuable catalysts for a wide variety of stereoselective Diels–Alder reactions. In a key step *en route* to

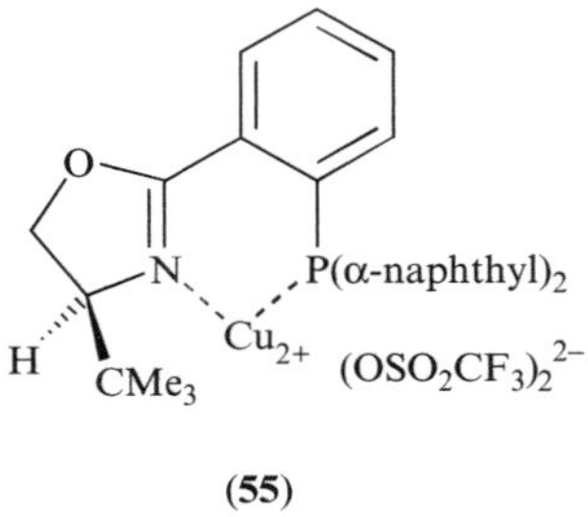

(55)

optically active cannabinols, Evans and co-workers showed that an acyclic dienol ester combined with the amide (**53**) to give the cyclohexane derivative (**54**) (Scheme 47)[135].

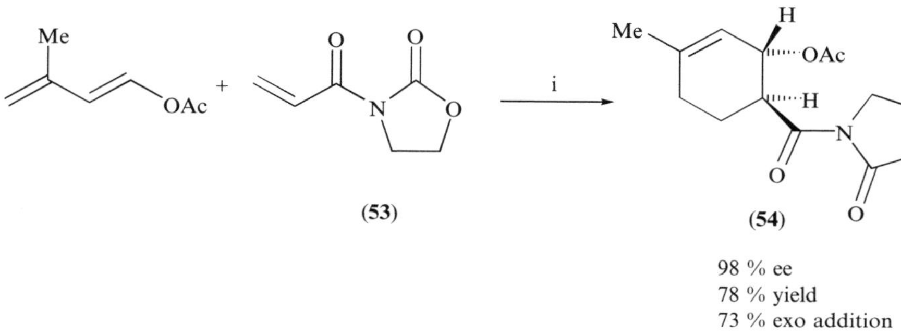
Me
OAc
+
O
O
N
O
i
Me
H
OAc
H
N
O
O
O
(53)
(54)
98 % ee
78 % yield
73 % exo addition

Scheme 47: Reagents and conditions: i) Compound (**48**) (2 mol%), CH_2Cl_2, $-20\,°C$.

The phosphino-oxazoline copper(II) complex (**55**) has also been found to be an effective catalyst[136] as have some titanium complexes, such as the extensively researched titanium–TADDOL system (**56**)[137]. A modified Ti(IV)–TADDOL compound is the catalyst of choice to promote Diels–Alder cycloaddition reactions between cyclopentadiene and alk-2-enyl phenylsulfonylmethyl ketones[138].

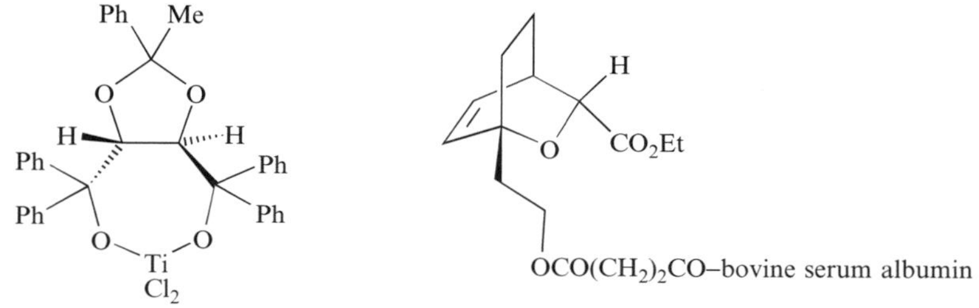
Ph
Me
O
O
H
H
Ph
Ph
Ph
Ph
O
O
Ti
Cl2
H
O
CO2Et
OCO(CH2)2CO–bovine serum albumin

(**56**) (**57**)

The library of natural catalysts has very little to offer for the catalysis of Diels–Alder (and the reverse) reactions (Diels–Alderases)[139]. For this reason one of the intriguing areas of biomimickry, namely the formation and use of antibodies exhibiting catalytic activity, has focused on [4 + 2] reactions to try to furnish proteins possessing useful catalytic properties. Thus in early studies a polyclonal catalytic antibody raised to hapten (**57**)[140] showed a modest rate enhancement for the reaction depicted in Scheme 48.

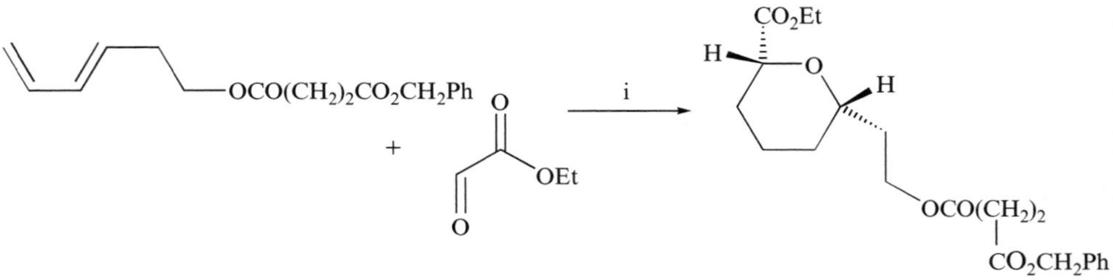

Scheme 48: <u>Reagents and conditions</u>: i) polyclonal antibody raised against hapten (**57**).

The same, understandable bias towards the preferred use of 'man-made' catalysts, rather than biocatalysts, continues in the area of hetero-Diels–Alder reactions[141]. For example, in the presence of 5 mol% of copper complexes of the type (**47**), cyclohexadiene and ethyl glyoxylate produce the oxabicyclooctene (**58**) (66% yield, 97% ee)[142].

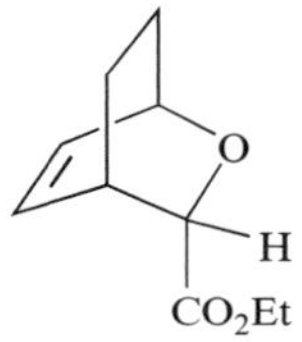

(**58**)

Asymmetric alkylation of enolates can be effected using chiral phase transfer reagents. In an example from O'Donnell's group, the ester (**59**) is alkylated in a two-phase solvent system containing an N-benzylcinchoninium salt (Scheme 49)[143]. Again, there is no competing methodology in the armoury of biocatalysis.

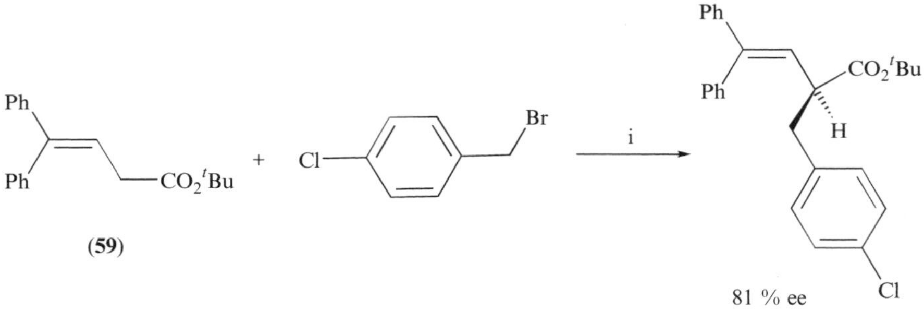

(**59**)

81 % ee

Scheme 49: <u>Reagents and conditions</u>: i) 50% NaOH, toluene and CH_2Cl_2, 5 °C N-benzylcinchoninium salt.

The promotion of carbon–carbon bond forming reactions involving alkenes is, once again, almost entirely within the domain of non-natural catalysts. One example each from five important areas are described below. It should be noted that this area is one of intense current interest and new catalysts and novel methodologies are appearing monthly; thus the following selection gives the reader only a glimpse of the important ground-breaking work in this area.

Hydrocyanation of alkenes (and alkynes) is an efficient route to nitriles *en route* to many types of fine chemicals. Initial studies of the hydrocyanation of vinylarenes such as styrene involved the use of a nickel–DIOP system, but ee's were disappointing at *ca* 10%. More success was achieved with carbohydrate derived phosphinite-nickel catalysts. For example the glucose-based bisphosphinite (60), on complexation with the metal, promoted the hydrocyanation of 4-methyl styrene to afford (*S*)-2-*para*-tolylpropanonitrile in 70% ee[144]. The same ligand promoted the asymmetric hydrocyanation of 2-methoxy-6-vinyl naphthalene to give an important intermediate to the nonsteroidal anti-inflammatory (NSAI) agent naproxen in *ca* 90% ee, using Ni(COD)$_2$ as the source of the metal. Also recently discovered has been a practical synthetic route to α-amino acids using titanium-catalysed enantioselective addition of cyanide to imines[145].

(60)   Ar = 3,5-bis-trifluorophenyl   (61)

Rhodium (I) complexes of chiral phosphines have been the archetypical catalysts for the hydrocarbonylation of 1-alkenes, with platinum complexes such as (61) making an impact also in the early 1990s[146]. More recently, rhodium(I)-chiral bisphosphites and phosphine–phosphinites have been investigated. Quite remarkable results have been obtained with Rh(I)–BINAPHOS (62), with excellent ee's being obtained for aldehydes derived for a wide variety of substrates[147]. For example, hydroformylation of styrene gave a high yield of (*R*)-2-phenylpropanal (94% ee). The same catalyst system promoted the conversion of *Z*-but-2-ene into (*S*)-2-methylbutanal (82% ee).

The related field involving the hydrocarboxylation of alkenes is also under investigation[148], not least because of its potential importance in the synthesis of NSAI drugs. An indirect way to the latter compounds involves the hydrovinylation of alkenes. For example catalysis of the reaction of ethylene with 2-methoxy-6-vinylnaphthalene at −70°C using (allylNiBr)$_2$ and binaphthyl (63)

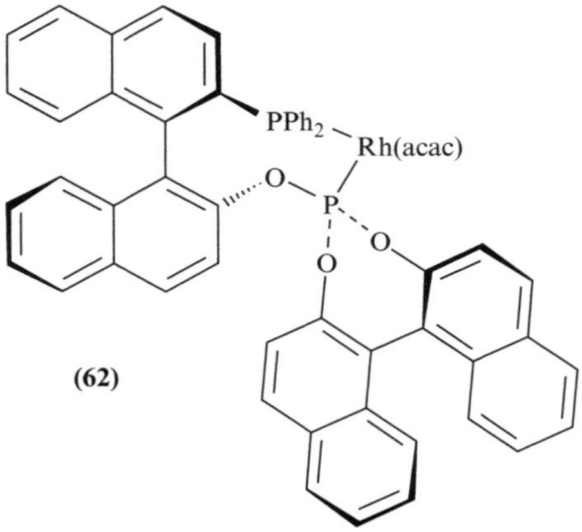

(62)

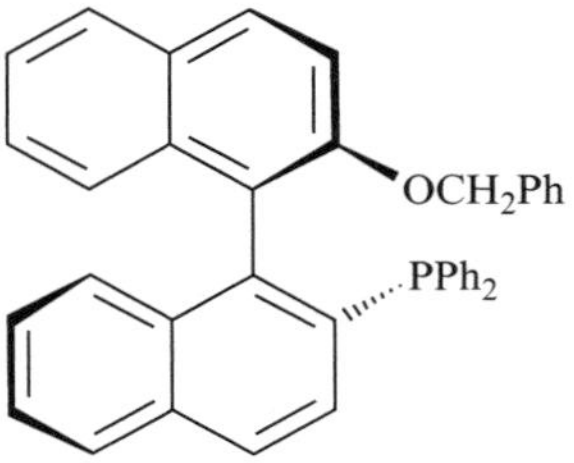

(63)

furnished the naproxen precursor (64) in 97% yield and 80% ee[149]. While nickel complexes have been most widely used for this type of process, palladium with a menthol-derived phosphinite has been used to convert ethene and styrene into (S)-3-phenylbut-1-ene in 66% yield and 86% ee[150].

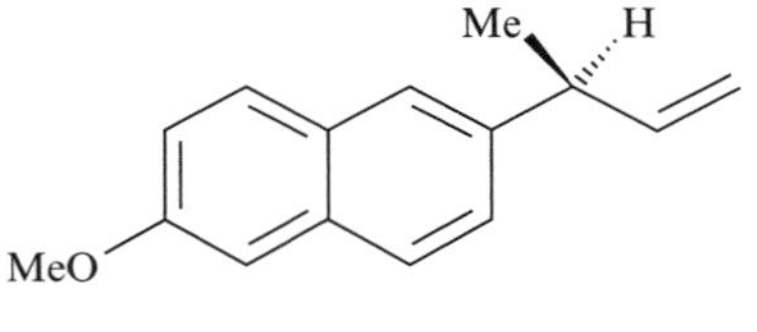

(64)

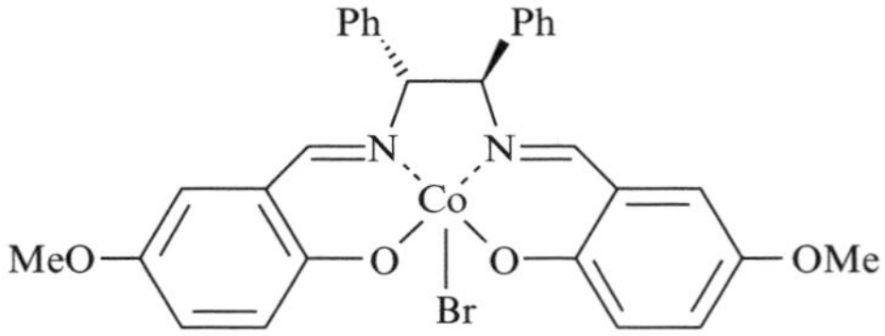

(65)

Cyclopropanation reactions can be promoted using copper or rhodium catalysts or indeed systems based on other metals. As early as 1965 Nozaki showed that chiral copper complexes could promote asymmetric addition of a carbenoid species (derived from a diazoester) to an alkene. This pioneering study was embroidered by Aratani and co-workers who showed a highly enantioselective process could be obtained by modifying the chiral copper

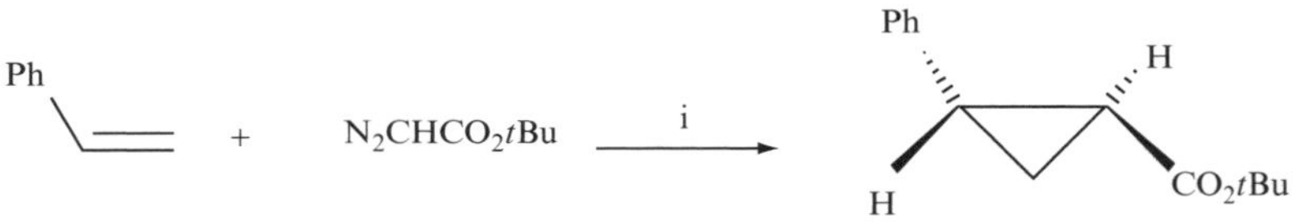

80 % (inc. 4 % cis)
93 % ee

Scheme 50: Reagents and conditions: (i) Catalyst (65), room temperature, CH_2Cl_2, 24 h.

complex[151]. Subsequently many excellent metal-catalysed methods have been developed for asymmetric cyclopropanation[152], most being trans-selective for the addition of diazo-ester to an alkene such as styrene: one example is shown in Scheme 50[153]. Only a few catalysts (for example a ruthenium–salen system) have been found that promote asymmetric cyclopropanation to give cis-products[154]. The range of asymmetric reactions of diazoesters has been extended to additions to imines to furnish aziridine derivatives[155].

Finally allylic substitution reactions involving, for example, replacement of an acetate unit with a malonate residue (or other nucleophiles) has been re-searched extensively by Trost and co-workers[156]. This group originally used Pd(PPh$_3$)$_4$ in the presence of a chiral phosphine to induce asymmetry but has shown more recently, inter alia, that the isomers (66) and (67) are both converted into the diester (68) in good yield and >95% ee using the dipyridine ligand (69) in a molybdenum-based catalyst (Scheme 51). The extensive range of chiral cata-lysts that have been used to effect enantioselective C–C and C–heteroatom bond formation is such allyl displacement reactions has been reviewed[157].

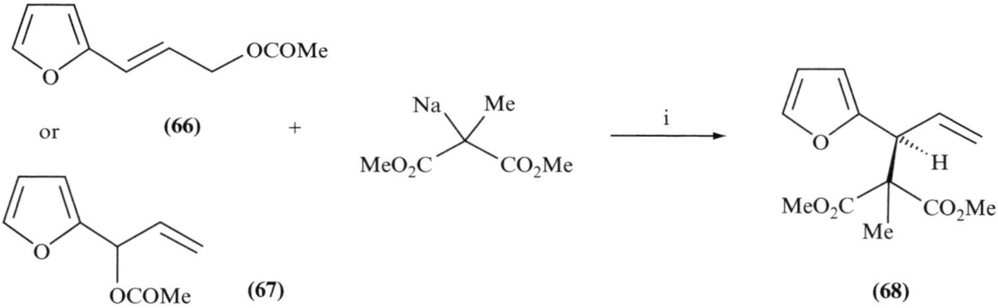

Scheme 51: <u>Reagents and conditions:</u> (i) 10% (MeCN$_3$Mo(CO)$_3$ ligand (69).

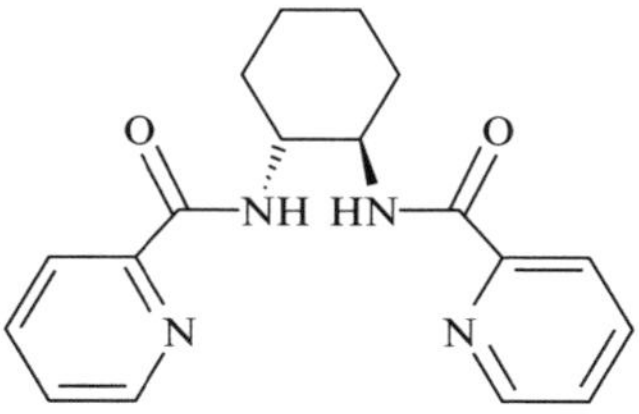

(69)

1.5 CONCLUSIONS

It is clear that in the following areas of synthetic chemistry the use of isolated enzymes or whole cell organisms should be considered (sometimes alongside other forms of catalysis) when one is faced with the transformation of the novel substrate.

- Enantioselective hydrolysis reactions, especially esters, amides and nitriles.
- Stereocontrolled oxidation of aromatic compounds (hydroxylation or dihydroxylation) and hydroxylation of some alicyclic compounds, especially at positions remote from pre-existing functionality.
- Stereocontrolled oxidation of sulfides to sulfoxides.
- Formation of optically active cyanohydrins.

Biomimetic reactions should also be considered for the preparation of optically active cyanohydrins (using a cyclic dipeptide as catalyst) and also for the epoxidation of α, β-unsaturated ketones (using polyleucine or congener as a catalyst).

In most other areas, especially in the field of carbon–carbon bond formation reactions, non-natural catalysts reign supreme.

However, while it is clear that biocatalysts may only provide viable and reliable methods in about 5–10% of all transformations of interest to synthetic organic chemists, it is also clear that in some cases the biotransformation will provide the **key step** in the best method in going from a cheap substrate to a high value, optically active fine chemical. Thus ignoring biotransformations altogether means one may occasionally overlook the best pathway to a target structure.

In addition there is at least one area where enzyme-catalysed reactions have established themselves as the first line of attack for solving synthetic problems; that area involves the transformations of carbohydrates. Indeed, biocatalysed transformations of saccharides is becoming increasingly popular and roughly 10% of the recent literature (Year 2000) on biotransformations involves the preparation and modification of carbohydrates. Early literature on chemoenzymatic approaches for the synthesis of saccharides and mimetics has been reviewed by a pioneer in the field, C.-H. Wong[158]. For one of the most popular areas, enzyme-catalysed glycosylation reactions, a useful survey is also available, penned by the same senior author[159].

One advantage of using enzyme-catalysed reactions in this field is that exquisite regio- and stereo-selectivity can be obtained, without recourse to long-winded protection/deprotection strategies. Furthermore, it is perfectly feasible to use different enzymes sequentially, quickly to produce complex polysaccharides. In the example shown in Scheme 50; N-acetylglucosamine is appended by a linker to a Sepharose bead: thereafter galactosyltransferase (with UDP-galactose), sialyltransferase (with CMP-neuraminic acid 5-acetate) and fucosyltransferase (with GDP-fucose) were used sequentially to prepare sialyl Lewis tetrasaccharide (**70**) attached to the solid support; an impressive overall yield of 57% was recorded[160].

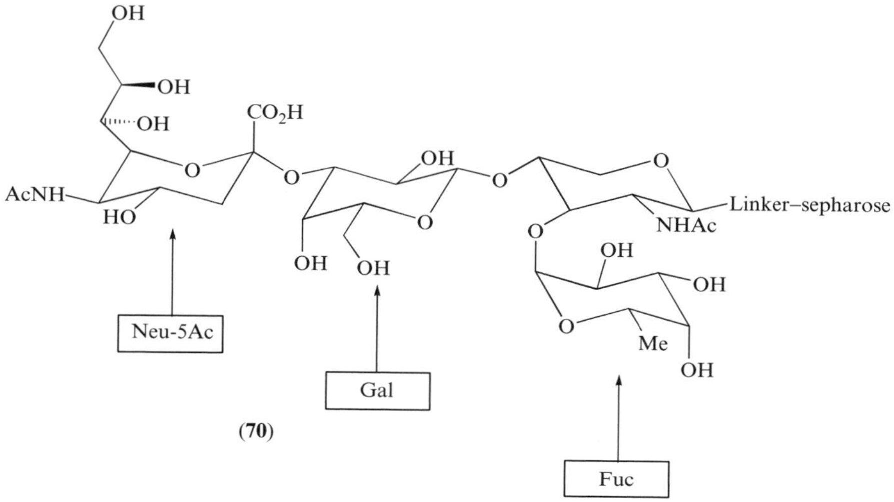

(70)

The sharp rise in the number of enzymes capable of promoting coupling reactions involving carbohydrate moieties mirrors the increased activity and interest in this field. Obviously this will provide an important niche area where enzyme-catalysed reactions will probably remain the methodology of choice at least for the foreseeable future.

So, in the final analysis, biocatalysis should not be considered in a separate sector available only to the specialist bioorganic chemist. It is one method, in the portfolio of catalytic techniques, that is available to all chemists for the solution of present and future problems in organic synthesis. To erect a 'Chinese wall' between the natural and non-natural catalysts is totally illogical and prevents some creative thinking, particularly in the area of coupled natural/ non-natural catalysts[161] and biomimetic systems[162].

REFERENCES

1. *Introduction to Biocatalysis using Enzymes and Microorganisms* by Roberts, S.M., Turner, N.J., Willetts, A.J. and Turner, M.K. Cambridge University Press, New York, 1995.
2. *Organic Synthesis with Oxidative Enzymes* by Holland, H.L., VCH, Weinheim, 1992.
3. An interesting snapshot of work on-going in the mid-1980s is found in the book *Biotransformations in Preparative Organic Chemistry* by Davies, H.G., Green, R.H., Kelly, D.R. and Roberts, S.M. Academic Press, London, 1989. For a modern work see *Biotransformations* by Faber, K. Springer Desktop Edition, 1999 or *Biocatalysis* by Fessner, W.-D. Springer Desktop Edition, 1999.
4. Klibanov, A.M. *Acc. Chem. Res.*, 1990, **23**, 114, A.M. Koskinen, P. and Klibanov, A.M. *Enzymatic Reactions in Organic Media*, Blackie Academic, London, 1996.
5. *Enzyme Catalysis in Organic Synthesis*, Volumes 1 and 2, eds Drauz, K.-H. and Waldmann, H., VCH, Weinheim, 1995.

6. A detailed review of the literature of non-enzymic catalysts is given in *Comprehensive Asymmetric Catalysis* eds Jacobsen, E.N., Pfaltz, A. and Yamamoto, H. Springer-Verlag, Berlin/Heidelberg, 1999. As an introductory text for post-graduate students see *Catalysis in Asymmetric Synthesis*, Williams, J.M.J. Sheffield Academic Press, Sheffield, UK, 1999. A comparison of biocatalysis versus chemical catalysis has also been made by Averill, B.A., Laane, N.W.M., Straathof, A.J.J. and Tramper, J., in *Catalysis: An Integrated Approach* (eds van Santen, R.A.; van Leeuwen, P.W.N.M., Moulijn, J.A. and Averill, B.A.) Elsevier, The Netherlands, 1999, Chapter 7.

7. $E = \ln[(1 - c)(1 - ee_s)]/\ln[(1 - c)(1 + ee_s)]$ where c = conversion ($100\% = 1.0$); ee_s = enantiomeric excess of substrate ($100\% = 1.0$).

8. Ozaki, E. and Sakashita, K. *Chem. Lett.*, 1997, 741.

9. Sakai, T., Miki, M., Nakatoni, M., Ema, T., Uneyama, K. and Utaka, M. *Tetrahedron Lett.*, 1998, **39** 5233.

10. Models of lipases and esterases, Jones, L.E. and Kazlauskas, R. *Tetrahedron: Asymmetry*, 1997, **8**, 3719; references in Schmid, R.D. and Verger, R. *Angew. Chem. Int. Ed* 1998, *37*, 1609; Colombo, G., Toba, S. and Merz, K.M. *J. Am. Chem. Soc.* 1999, **121**, 3486.

11. Bakke, M., Takizawa, M., Sugai, T. and Ohta, H. *J. Org. Chem.* 1998, **63**, 6929.

12. Guanti, G., Narisano, E. and Riva, R. *Tetrahedron: Asymmetry*, 1998, **9**, 1859.

13. Bonini, C., Giugliano, A., Racioppi, R. and Righi, G. *Tetrahedron Lett.*, 1996, **37** 2487.

14. *Hydrolases in Organic Synthesis* by Bornscheuer, U.T. and Kazlauskas, R.J. Wiley-VCH, Weinheim, 1999.

15. Luna, A., Astorga, C., Fülöp, F. and Gotor, V. *Tetrahedron: Asymmetry*, 1998, **9**, 4483.

16. Fujita, T., Tanaka, M., Norimine, Y., Suemune, H. and Sakai, K. *J. Org. Chem.*, 1997, **62**, 3824.

17. Copeland, G.T., Jarvo, E.R. and Miller, S.J. *J. Org. Chem.*, 1998, **63**, 6784.

18. Oriyama, T., Imai, K., Hosoya, T. and Sano, T. *Tetrahedron Lett.*, 1998, **39**, 397.

19. Vedejs, E. and Daugulis, O. *J. Am. Chem. Soc.*, 1999, **121**, 5813; Tao, B., Ruble, I.C., Hoic, D.A. and Fu, G.C. ibid, 1999, **121**, 5091; Jarvo, E.R., Copeland, G.T., Papaioannou, N., Bonitatebus, P.J., Jr. and Miller, S.J. *ibid*, 1999, **121**, 11638.

20. Effenberger, F., Graef, B.W. and Osswald, S. *Tetrahedron Asymmetry*, 1997, **8**, 2749.

21. Taylor, S.J.C., Brown, R.C., Keene, P.A. and Taylor, I.N. *Bioorg. Med. Chem.*, 1999, **7**, 2163.

22. Faber, K., Mischitz, M. and Kroutil, W. *Acta Chem. Scand.*, 1996, **50**, 249; Archelas, A. and Furstoss, R. *Ann. Rev. Microbiol.*, 1997, **51**, 491; see also C.A.G.M. Weijers, *Tetrahedron: Asymmetry*, 1997, **8**, 639.

23. Pedragosa-Moreau, S., Archelas, A. and Furstoss, R. *Tetrahedron Lett.*, 1996, **37**, 3319.

24. Pedragosa-Moreau, S., Archelas, A. and Furstoss, R. *Tetrahedron*, 1996, **52**, 4593; Pedragosa-Moreau, S., Morisseau, C., Zylber, J., Archelas, A., Baratti, J. and Furstoss, R. *J. Org. Chem.*, 1996, **61**, 7402. A theoretical analysis of such epoxide ring-opening reactions has been published, Moussou, P., Archelas, A., Baratti, J. and Furstoss, R. *Tetrahedron: Asymmetry*, 1998, **9**, 1539.

25. A review of the present position is available, Orru, R.V.A., Archelas, A., Furstoss, R. and Faber, K. *Adv. Biochem. Eng. Biotechnol.* 1999, **63**, 146; note that the

epoxide hydrolase from *Agrobacterium radiobacter* has been over-expressed – Spelberg, J.H.L., Rink, R., Kellogg, R.M. and Janssen, D.B. *Tetrahedron: Asymmetry*, 1998, **9**, 459.

26. Jacobsen, E.N., Katiuchi, F., Konsler, R.G., Larrow, J.F. and Tokunaga, M. *Tetrahedron Lett.*, 1997, **38**, 773; Annis, D.A. and Jacobsen, E.N. *J. Am. Chem. Soc.*, 1999, **121**, 4147.
27. Maddrell, S.J., Turner, N.J., Kerridge, A., Willetts, A.J. and Crosby, J. *Tetrahedron Lett.*, 1996, **37**, 6001.
28. Kreutz, O.C., Moran, P.J.S. and Rodrigues, J.A.R. *Tetrahedron: Asymmetry*, 1997, **8**, 2649.
29. Sugai, T., Hamada, K., Akeboshi, T., Ikeda, H. and Ohta, H. *Synlett*, 1997, 983.
30. Recently bakers' yeast reductions in petroleum ether has been explored, Medson, C., Smallridge, A.J. and Trewella, M.A. *Tetrahedron: Asymmetry*, 1997, **8**, 1049.
31. For a brief survey of the background to (and nomenclature in) whole-cell biotransformations see *Introduction to Biocatalysis using Enzymes and Micro-organisms* by Roberts, S.M., Turner, N.J., Willetts, A.J. and Turner, M.K. Cambridge University Press, New York, 1995, Chapter 2, p. 34–78.
32. Cui, J.-N., Ema, T., Sakai, T. and Utaka, M. *Tetrahedron: Asymmetry*, 1998, **9**, 2681; Dao, D.H., Okamura, M., Akasaka, T., Kawai, Y., Hida, K. and Ohno, A. *Tetrahedron: Asymmetry*, 1998, **9**, 2725; Hayakawa, R., Nozawa, K., Shimizu, M. and Fujisawa, T. *Tetrahedron Lett.*, 1998, **39**, 67.
33. Seelbach, K., Riebel, B., Hummel, W., Kula, M.-R., Tishkov, V.I., Egorov, A.M., Wandrey, C. and Kragl, U. *Tetrahedron Lett.*, 1996, **37**, 1377.
34. Bakos, J., Tōth, D., Heil, B. and Markō, L. *J. Organomet. Chem.*, 1985, **279**, 23.
35. Ohkuma, T., Ooka, H., Hashigushi, S., Ikariya, T. and Noyori, R. *J. Am. Chem. Soc.*, 1995, **117**, 2675.
36. Doucet, H., Ohkuma, T., Murata, K., Yokozawa, T., Kozawa, M., Katayama, E., England, A.F., Ikariya, T. and Noyori, R. *Angew. Chem., Int. Ed., Engl.*, 1998, **37**, 1703; Ohkuma, T., Koizumi, M., Doucet, H., Pham, T., Kogawa, M., Murata, K., Katayama, E., Yokazawa, T., Ikarija, T. and Noyori, R. *J. Am. Chem. Soc.*, 1998, **120**, 13529.
37. Zhang, X., Taketomi, T., Yoshizumi, T., Kumobayashi, H., Akutagawa, S., Mashima, K. and Takaya, H. *J. Am. Chem. Soc.*, 1993, **115**, 3318.
38. Palmer, M.J. and Wills, M. *Tetrahedron Asymmetry*, 1999, **10**, 2045; Jiang, Y., Jiang, Q. and Zhang, X. *J. Am. Chem. Soc.*, 1998, *120*, 3817; de Bellefon, C. and Tanchoux, N. *Tetrahedron Asymmetry* 1998, **9**, 3677.
39. Corey, E.J. and O Link, J. *Tetrahedron Lett.*, 1989, **30**, 6275; Corey, E.J. and Helal, C.J. *Angew. Chem. Int. Ed. Engl.*, 1998, **37**, 1986.
40. Mathre, D.J., Thompson, A.S., Douglas, A.W., Hoogsteen, K., Carroll, J.D., Corley, E.G. and Grabowski, E.J.J. *J. Org. Chem.*, 1993, **58**, 2880.
41. Salunkhe, A.M. and Burkhardt, E.R. *Tetrahedron Lett.*, 1997, **38**, 1523.
42. Zhu, Y.Y. and Burnell, D.J. *Tetrahedron: Asymmetry*, 1996, **7**, 3295.
43. Crocque, V., Masson, C., Winter, J., Richard, C., Lemaitre, G., Lenay, J., Vivat, M., Buendia, J. and Pratt, D. *Org. Process Res. Dev.*, 1997, **1**, 2.
44. Seebach, D., Sutter, M.A., Weber, R.H. and Züger, M.F. *Org. Synth.*, 1984, **63**, 1.
45. Dahl, A.C. and Madsen, J.O. *Tetrahedron: Asymmetry*, 1998, **9**, 4395.
46. Noyori, R., Ohkuma, T., Kitamura, M., Takaya, H., Sayo, N., Kumobayashi, H. and Akutagawa, S. *J. Am. Chem. Soc.*, 1987, **109**, 5856.

47. Tas, D., Thoelen, C., Vanekelecom, I.F.J. and Jacobs, P.A. *J.C.S., Chem. Commun.*, 1997, 2323.
48. Kitamura, M., Ohkuma, T., Inoue, S., Sayo, N., Kumobayashi, H., Akutagawa, S., Ohta, T., Takayon, H. and Norori, R. *J. Am. Chem. Soc.*, 1988, **110**, 629.
49. Kawano, H., Ishii, Y., Saburi, M. and Uchida, Y. *J.C.S. Chem. Commun.*, 1988, 87.
50. Arrigo, P.D., Fuganti, C., Fantoni, G.P. and Servi, S. *Tetrahedron*, 1998, **54**, 15017.
51. Aleu, J., Fronza, G., Fuganti, C., Perozzo, V. and Serra, S. *Tetrahedron: Asymmetry*, 1998, **9**, 1589.
52. Kawai, Y., Haynshi, M., Inaba, Y., Saitou, K. and Ohno, A. *Tetrahedron Lett.*, 1998, **39**, 5225; Kawai, Y., Saitou, K., Hida, K., Dao, D.H. and Ohno, A. *Bull. Chem. Soc., Japan*, 1996, **69**, 2633.
53. Ferraboschi, P., Elahi, S.R.T., Verza, E., Meroni-Rivolla, F. and Santaniello, E. *Synlett*, 1996, 1176.
54. Leuenberger, H.G.W., Boguth, W., Barner, R., Schmid, M. and Zell, R. *Helv. Chim. Acta.*, 1979, **62**, 455.
55. Matsumoto, K., Kawabata, Y., Takahashi, J., Fujita, Y. and Hatanaka, M. *Chem. Lett.*, 1998, 283.
56. *Homogeneous Catalysis* Parshall, G.W. and Ittel, S.D. (eds), 1992 (second edition), Wiley, New York, p 33; Takaya, H., Ohta, T. and Noyori, R. *Catalytic Asymmetric Synthesis* Chapter 1, VCH, Weinheim, 1993.
57. Ikariya, T., Ishii, Y., Kawano, H., Arai, T., Saburi, M., Yoshikawa, S. and Akutagawa, S. *J.C.S., Chem. Commun.*, 1985, 922.
58. Noyori, R. *Asymmetric Catalysis in Organic Synthesis*, Wiley, New York, 1994.
59. Kitamura, M., Hsiao, Y., Noyori, R. and Takaya, H. *Tetrahedron Lett.*, 1987, **28**, 4829.
60. Kitamura, M., Hsiao, Y., Ohta, M., Tsukamoto, M., Ohta, T., Takaya, H. and Noyori, R. *J. Org. Chem.*, 1994, **59**, 297.
61. Uemura, T., Zhang, X.Y., Matsumura, K., Sayo, N., Kumobayashi, H., Ohta, T., Nozaki, K. and Takaya, H. *J. Org. Chem.*, 1996, **61**, 5510; a classical example is the Monsanto synthesis of (L)-DOPA using ruthenium complexed with the diphosphine DIPAMP, see *Classics in Total Synthesis*, Nicolaou, K.C. and Sorensen, E.J. VCH, Weinheim, 1996.
62. Dwars, T., Schmidt, U., Fischer, C., Grassert, I., Kempe, R., Frölich, R., Drauz, K. and Oehme, G. *Angew. Chem. Int. Ed*, 1998, **37**, 2851.
63. Burk, M.J., Gross, M.F. and Martinez, J.P. *J. Am. Chem. Soc.*, 1995, **117**, 9375; see also Burk, M.J., Allen, J.G. and Kiesman, W.F. *J. Am. Chem. Soc.*, 1998, **120**, 657; for the asymmetric reduction of a variety of β-substituted esters using rhodium/ligand complexes of this type see Burk, M.J., Bienewald, F., Harris, M. and Zanotti-Gerosa, A. *Angew Chem. Int. Ed. Engl.*, 1998, **37**, 1931.
64. Boaz, N.W. *Tetrahedron Lett.*, 1998, **39**, 5505; see also Zhu, G., Casalnuova, A.L. and Zhang, X. *J. Org. Chem.*, 1998, **63**, 8100.
65. Reetz, M.T., Gosberg, A., Goddard, R. and Kyung, S.-H. *J.C.S. Chem. Comm.* 1998, 2077; Kang, J., Lee, J.H., Ahn, S.H. and Choi, J.S. *Tetrahedron Lett.*, 1998, **39**, 5523; Perea, J.J.A., Borneo, A. and Knochel, P. *ibid*, 1998, **39**, 8073.
66. Chan, A., Hu, W.H., Pai, C.C., Lau, C.P., Jiang, Y.Z., Mi, A.Q., Yan, M., Sun, J., Lou, R.L. and Deng, J.G. *J. Am. Chem. Soc.*, 1997, **119**, 9570.
67. Rajan Babu, T.V., Ayers, T.A., Halliday, G.A., You, K.K. and Calabrese, J.C. *J. Org. Chem.*, 1997, **62**, 6012; Selke, R., Ohff, M. and Riepe, A. *Tetrahedron*, 1996, **52**, 15079.

68. Derrien, N., Dousson, C.B., Roberts, S.M., Berens, U., Burk, M.J. and Ohff, M. *Tetrahedron: Asymmetry*, 1999, **10**, 3341.
69. Doi, T., Kokubo, M., Yamamoto, K. and Takahashi, T. *J. Org. Chem.*, 1998, **63**, 428.
70. Cyclohexene and simple derivatives may be oxidized in the allylic position with a fair degree of stereocontrol using non-natural catalysts, see for example Schulz, M., Kluge, R. and Gelacha, F.G. *Tetrahedron Asymmetry*, 1998, **9**, 4341.
71. For a review of early work see *Biotransformations in Preparative Organic Chemistry*, Davies, H.G., Green, R.H., Kelly, D.R. and Roberts, S.M. Academic Press, London, 1989.
72. *Medical Chemistry: the Role of Organic Chemistry in Drug Research* eds Price, B.J. and Roberts, S.M. Academic, Orlando, 1985.
73. Hu, S., Tian, X., Zhu, W. and Fang, Q. *Tetrahedron*, 1996, **52**, 8739; Hu, S., Sun, S., Tian, X. and Fang, Q. *Tetrahedron Lett.*, 1997, **38**, 2721.
74. Aitken, S.J., Grogan, G., Chow, C.S.-Y., Turner, N.J. and Flitsch, S.L. *J.C.S. Perkin Trans. I*, 1998, 3365.
75. Palmer, C.F., Webb, B., Broad, S., Casson, S., McCague, R., Willetts, A.J. and Roberts, S.M. *Bioorg. Med. Chem. Lett.*, 1997, **7**, 1299.
76. Jones, M.E., England, P.A., Rouch, D.A. and Wong, L.-L. *J.C.S., Chem. Commun.*, 1996, 2413; England, P.A., Rough, D.A., Westlake, A.C.G., Bell, E.G., Nickerson, D.P., Webberley, M., Flitsch, S.L. and Wong, L.L. *J.C.S., Chem. Commun.*, 1996, 357.
77. de Raadt, A., Griengl, H., Petsch, M., Plachota, P., Schoo, N., Weber, H., Braunegg, G., Kopper, I., Kreiner, M., Zeiser, A. and (in part) Kieslich, K. *Tetrahedron: Asymmetry*, 1996, **7**, 467, 473, 491.
78. Torimura, H., Yoshida, H., Kano, K., Ikeda, T., Nagaswa, T. and Ueda, T. *Chem. Lett.*, 1998, 295; Kulla, H.G. *Chimia*, 1991, **45**, 51.
79. Dingler, C., Ladner, W., Krei, G.A., Cooper, B. and Hauer, B. *Pesticide Sci.*, 1996, **46**, 33.
80. For recent information of other dienediols prepared by this method and for the range of products prepared from these compounds see Roberts, S.M. *J.C.S. Perkin Trans. 1.*, 1998, 164; 1999, 10; 2000, 623.
81. Sharpless, K.B., Amberg, W., Bennani, Y.L., Crispino, G.A., Hartung, J., Jeong, K.-S., Kwong, H.-L., Morikawa, K., Wang, Z.-M., Xu, D. and Zhang, X.-L. *J. Org. Chem.*, 1992, **57**, 2768; Kolb, H.C., VanNieuwenhze, M.S. and Sharpless, K.B. *Chem. Rev.*, 1994, **94**, 2483.
82. Markõ, I.E. and Svendsen, J.S. in *Comprehensive Organometallic Chemistry* (ed. L.S. Hegedus) Vol. 12, p. 1137, Pergamon, Oxford, 1995.
83. Crispino, G. and Sharpless, K.B. *Synlett*, 1993, 47; Nambu, M. and White, J.D. *J.C.S., Chem. Commun.*, 1996, 1619; Takano, S., Yoshimitsu, T. and Ogasawara, K. *J. Org. Chem.*, 1994, **59**, 54;
84. Reddy, K.L., Dress, K.R. and Sharpless, K.B. *Tetrahedron Lett.*, 1998, **39**, 3667; P. O'Brien, *Angew. Chem. Int. Ed. Engl.*, 1999, **38**, 326.
85. Schweiter, M.J. and Sharpless, K.B. *Tetrahedron Lett.*, 1985, **26**, 2543.
86. Carlier, P.R., Mungall, W.S., Schroder, G. and Sharpless, K.B. *J. Am. Chem. Soc.*, 1988, **110**, 2978.
87. Yamamoto, H. and Oritani, T. *Biosci. Biotech. Biochem.*, 1994, **58**, 992.
88. Katsuki, T. *J. Mol. Cat.*, 1996, **113**, 87; Jacobsen, E.N. in *Comprehensive Organometallic Chemistry*, (eds Wilkinson, G., Stone, F.G.A., Abel, R.W. and Hegedus, L.S.), Pergamon. New York, 1995, Chapter 11.1.

89. Larrow, J.F. and Jacobsen, E.N. *Org. Synth.*, 1997, **75**, 1.
90. Larrow, J.F. and Jacobsen, E.N. *Org. Synth.*, 1998, **76**, 46.
91. Pietkainen, P. *Tetrahedron*, 1998, **54**, 4319.
92. Collman, J.P., Wang, Z., Straumanis, A. and Quelquejeu, M. *J. Am. Chem. Soc.*, 1999, **121**, 460.
93. Yang, D., Yip, Y.-C., Tang, M.-W., Wong, M.-K., Zheng, J.-H. and Cheung, K.-K. *J. Am. Chem. Soc.*, 1996, **118**, 491.
94. Wang, Z.-X., Tu, Y., Frohn, M. and Shi, Y. *J. Org. Chem.*, 1997, **62**, 2328; Wang, Z.-X., Tu, Y., Frohn, M., Zhang, J.-R. and Shi, Y. *J. Am. Chem. Soc.*, 1997, **119**, 11224; Zhu, Y., Tu, Y., Yu, H. and Shi, Y. *Tetrahedron Lett.* 1998, **39**, 7819; Shi, Y. and Shu, L. *ibid*, 1999, **40**, 8721.
95. Armstrong, A. and Hayter, B.R. *J.C.S., Chem. Commun.*, 1998, 621; *idem, Tetrahedron*, 1999, **55**, 11119.
96. Enders, D., Zhu, J. and Kramps, L. *Liebigs Ann. Recueil*, 1997, 1101.
97. Elston, C.L., Jackson, R.F.W., MacDonald, S.J.F. and Murray, P.J. *Angew. Chem. Int. Ed. Engl.*, 1997, **36**, 410.
98. Lygo, B. and Wainwright, P.G. *Tetrahedron*, 1999, **55**, 6289.
99. For a comparison of all the methodologies in this area, see Porter, M.J. and Skidmore, J. *J.C.S. Chem. Commun.*, 2000, 1215.
100. Bougauchi, M., Watanabe, S., Arai, T., Sasai, H. and Shibasaki, M. *J. Am. Chem. Soc.*, 1997, **119**, 2329; such epoxidations can benefit from the addition of Ph$_3$PO, see Daikai, K., Kamaura, M. and Inanaga, J. *Tetrahedron Lett.*, 1998, **39**, 7321.
101. Porter, M., Roberts, S.M. and Skidmore, J. *Bioorg. Med. Chem.*, 1999, **8**, 2145.
102. Mazzini, C., Lebreton, J., Alphand, V. and Furstoss, R. *J. Org. Chem.*, 1997, **62**, 5215 and references therein.
103. For a full review on enzyme-catalysed Baeyer–Villiger oxidations see Roberts, S.M. and Wan, P.W.H. *J. Mol. Cat. B. Enzymatic*, 1998, **4**, 111.
104. Stewart, J.D., Reed, K.W., Martinez, C.A., Zhu, J., Chen, G. and Kayser, M.M. *J. Am. Chem. Soc.*, 1998, **120**, 3541.
105. Bolm, C., Luong, T.K.K. and Schlingloff, G. *Synlett*, 1997, 1151.
106. Lopp, M., Paju, A., Kanger, T. and Pehk, T. *Tetrahedron Lett.*, 1996, **37**, 7583; see also Kanger, T., Kriis, K., Paju, A., Pekk, T. and Lopp, M. *Tetrahedron Asymmetry*, 1998, **9**, 4475.
107. M.P.J., van Deurzen, vanRantwijk, F. and Sheldon, R.A. *Tetrahedron*, 1997, **53**, 13183.
108. Colonna, S., Gaggero, N., Carrea, G. and Pasta, P. *J.C.S., Chem. Commun.*, 1997, 439.
109. Yamanoi, Y. and Imamoto, T. *J. Org. Chem.*, 1997, **62**, 8560.
110. Reetz, M.T., Merk, C., Naberfeld, G., Rudolph, J., Griebenow, N. and Goddard, R. *Tetrahedron Lett.*, 1997, **38**, 5273; see also Superchi, S., Donnoli, M.I. and Rosini, C. *Tetrahedron Lett.* 1998, **39**, 8541.
111. Bolm, C., Schlingloff, G. and Bienewald, F. *J. Mol. Cat.*, 1997, **117**, 347.
112. Gregory, R.H.J. *Chem. Rev.*, 1999, **99**, 3649.
113. Han, S., Lin, G. and Li, Z. *Tetrahedron: Asymmetry*, 1998, **9**, 1935.
114. Griengl, H., Klempier, N., Pöchlauer, P., Schmidt, M., Shi, N. and Zabinskaja-Mackova, A.A. *Tetrahedron*, 1998, **54**, 14477.
115. Förster, S., Roos, J., Effenberger, F., Wajant, H. and Spauer, A. *Angew. Chem. Int. Ed. Engl.*, 1996, **35**, 437.

116. Hulst, R., Broxterman, Q.B., Kamphuis, J., Formaggio, F., Crisma, M., Toniolo, C. and Kellogg, R.M. *Tetrahedron: Asymmetry*, 1997, **8**, 1987; Shvo, Y., Gal, M., Becker, Y. and Elgavi, A. *Tetrahedron: Asymmetry*, 1996, **7**, 911; Kogut, E., Thoen, J.C. and Lipton, M.A. *J. Org. Chem.*, 1998, **63**, 4604; M. North, *Synlett*, 1993, 807.

117. Nitta, H., Yu, D., Kudo, M. and Inoue, S. *J. Am. Chem. Soc.*, 1992, **114**, 7969; Abe, H., Nitta, H., Mori, A. and Inoue, S. *Chem. Lett.*, 1992, 2443.

118. Belokon, Y.N., Green, B., Ikonnikov, N.S., North, M. and Tararov, V.I. *Tetrahedron Lett.*, 1999, **40**, 8147.

119. Abiko, A. and Wang, G. *J. Org. Chem.*, 1996, **61**, 2264.

120. Hamashima, Y., Sawada, D., Kanai, M. and Shibasaki, M. *J. Am. Chem. Soc.*, 1999, **121**, 2641.

121. Takayama, S., Martin, R., Wu, J., Laslo, K., Siuzda, G. and Wong, C.-H. *J. Am. Chem. Soc.*, 1997, **119**, 8146.

122. Kimura, T., Vassilev, V.P., Shen, G.-J. and Wong, C.-H. *J. Am. Chem. Soc.*, 1997, **119**, 11734.

123. Kobayashi, S., Fujishita, Y. and Mukaiyama, T. *Chem. Lett.*, 1990, 1455.

124. Parmee, E.R., Hong, Y.P., Tempkin, O. and Masamune, S. *Tetrahedron Lett.*, 1992, **33**, 1729.

125. Carreira, E.M., Singer, R.A. and Lee, W.S. *J. Am. Chem. Soc.*, 1994, **116**, 8837.

126. Evans, D.A., Kozlowski, M.C., Burgey, C.S. and MacMillan, D.W.C. *J. Am. Chem. Soc.*, 1997, **119**, 7893; see also Ghosh, A.K., Mathivanan, P. and Cappiello, J. *Tetrahedron Lett.*, 1997, **38**, 2427.

127. Evans, D.A., Rovis, T., Kozlowski, M.C. and Tedrow, J.S. *J. Am. Chem. Soc.*, 1999, **121**, 1994.

128. Yoshikawa, N., Yamada, Y.M.A., Das, J., Sasai, H. and Shibasaki, M. *J. Am. Chem. Soc.*, 1999, **121**, 4168.

129. An excellent overview of the stereochemistry of the aldol reaction is given by Procter, G. in *Asymmetric Synthesis*, Chapter 5, pp. 69–101, OUP, Oxford, 1996.

130. Corey, E.J., Sarshar, S. and Lee, D.-H. *J. Am. Chem. Soc.*, 1994, **116**, 12089.

131. Heller, D.P., Goldberg, D.R. and Wulff, W.D. *J. Am. Chem. Soc.*, 1997, **119**, 10551.

132. Corey, E.J. and Loh, T.-P. *J. Am. Chem. Soc.*, 1991, **113**, 8966.

133. Kamahori, K., Ito, K. and Itsuno, S. *J. Org. Chem.*, 1996, **61**, 8321.

134. Ishihara, K., Kondo, S., Kurihara, H. and Yamamoto, H. *J. Org. Chem.*, 1997, **62**, 3026.

135. Evans, D.A., Shaughnessy, E.A. and Barnes, D.M. *Tetrahedron Lett.*, 1997, **38**, 3193.

136. Sagasser, I. and Helmchen, G. *Tetrahedron Lett.*, 1998, **39**, 261.

137. Yamamoto, I. and Narasaka, K. *Chem. Lett.*, 1995, 1129.

138. Wada, E., Pei, W. and Kanemasa, S. *Chem. Lett.*, 1994, 2345.

139. Laschat, S. *Angew. Chem. Int. Ed. Engl.*, 1996, **35**, 289; for details of an interesting RNA Diels–Alderase see Tarasow, T.M., Tarasow, S.L., Tu, C., Kellogg, E. and Eaton, B.E. *J. Am. Chem. Soc.*, 1999, **121**, 3614.

140. Hu, Y.-J., Ji, Y.-Y., Wu, Y.-L., Yang, B.H. and Yeh, M. *Bioorg. Med. Chem. Lett.*, 1997, **7**, 1601.

141. Aza-Diels–Alder reactions (e.g. Yao, S., Johannnsen, M., Hazell, R.G. and Jorgensen, K.A. *Angew. Chem. Int. Ed.* 1998, **37**, 3121; Bromidge, S., Wilson, P.C. and Whiting, A. *Tetrahedron Lett.*, 1998, **39**, 8905) and oxa-Diels–Alder reactions (e.g.

Schaus, S.E., Branalt, J. and Jacobsen, E.N. *J. Org. Chem.*, 1998, **63**, 403) can be catalysed using chiral organometallic systems.

142. Johannsen, M. and Jogensen, K.A. *Tetrahedron*, 1996, **52**, 7321. (The organochromium catalysts invented by Jacobsen are also noteworthy, see Thompson, C.F., Jamison, T.F. and Jacobsen, E.N. *J. Am. Chem. Soc.*, 2000, **122**, 10482.)

143. Esikova, I.A., Nahreini, T.S. and O'Donnell, M.J. in *Phase-Transfer Catalysis* (M. Halpern, ed.), ACS (ACS Symposium Series), Washington, 1997, pp 89–96.

144. a) Casalnuovo, A.L. and Rajan Babu, T.V. *J. Am. Chem. Soc.*, 1994, **116**, 9869; *idem*, 'The Asymmetric Hydrocyanation of Vinyl Arenes' in *Chirality and Industry II* (eds Collins, A.N., Sheldrake, G.N. and Crosby, J.) Wiley, New York, 309.

145. Krueger, C.A., Kuntz, K.W., Drierba, C.D., Wirschun, W.G., Gleason, J.D., Snapper, M.L. and Hoveyda, A.H. *J. Am. Chem. Soc.*, 1999, **121**, 4284.

146. Agbossou, F., Carpentier, J.-F. and Mortreux, A. *Chem. Rev.*, 1995, **95**, 2485; Herrmann, W.A. and Cornils, B. *Angew. Chem. Int. Ed. Eng.*, 1997, **36**, 1048; I. Tôth, Elsevier, C.J., de Vries, J.G., Bakos, J., Smeets, W.J.J. and Spek, A.L. *J. Organomet. Chem.*, 1997, **540**, 15.

147. Sakai, N., Mano, S., Nozaki, K. and Takaya, H. *J. Am. Chem. Soc.*, 1993, **115**, 7033; Nozaki, K., Sakai, N., Nanno, T., Higashijima, T., Mano, S., Horiuchi, T. and Takaya, H. *ibid*, 1997, **119**, 4413.

148. Alper, H. and Hamel, N. *J. Am. Chem. Soc.*, 1990, **112**, 2803; Zhou, H., Hou, J., Chen, J., Lu, S., Fu, H. and Wang, H. *J. Organomet. Chem.*, 1997, **543**, 227.

149. Nomura, N., Jin, J., Park, H. and Rajan Babu, T.V. *J. Am. Chem. Soc.*, 1998, **120**, 459.

150. Bayersdörfer, R., Ganter, B., Englert, U., Keim, W. and Vogt, D. *J. Organomet. Chem.*, 1998, **552**, 187.

151. Aratani, T. *Pure Appl. Chem.*, 1985, **57**, 1839.

152. Doyle, M.P., McKervey, M.A. and Ye, T. *Modern Catalytic Methods for Organic Synthesis with Diazo Compounds*, John Wiley and Sons, New York, 1998; Lo, M.M-C. and Fu, G.C. *J. Am. Chem. Soc.* 1998, **120**, 10270.

153. Fukuda, T. and Katsuki, T. *Tetrahedron*, 1997, **53**, 7201.

154. Ishitani, H. and Achiwa, K. *Synlett*, 1997, 781; Uchida, T., Irie, R. and Katsuki, T. *ibid*, 1999, 1163; Niimi, T., Uchida, T., Irie, R. and Katsuki, T. *Tetrahedron Lett.*, 2000 **41**, 3647.

155. Antilla, J.C. and Wulff, W.D. *J. Am. Chem. Soc.*, 1999, **121**, 5099.

156. Trost, B.M. and Hachiya, I. *J. Am. Chem. Soc.*, 1998, **120**, 1104.

157. Trost, B.M. and van Vranken D.L. *Chem. Rev.*, 1996, **96**, 395; Trost., B.M. *Acc. Chem. Res.*, 1996, **29**, 355; Tye, H. *J.C.S. Perkin Trans. 1*, 2000, 284; see also Hamada, Y., Seto, N., Takayanagi, Y., Nakano, T. and Hara, O. *Tetrahedron Lett.*, 1999, **40**, 7791.

158. Sears, P. and Wong, C.-H. *J.C.S., Chem. Commun.*, 1998, 1161.

159. Takayama, S., McGarvey, G.J. and Wong, C.-H. *Chem. Soc. Rev.*, 1997, **26**, 407.

160. Blixt, O. and Norberg, T. *J. Org. Chem.*, 1998, **63**, 2705.

161. The work of Williams and Bäckvall provide a foretaste of possibilities in this general area see Persson, B.A., Larsson, A.L.E., Le Ray, M. and Bäckvall, J.-E. *J. Am. Chem. Soc.*, 1999, **121**, 1645; Allen, J.V. and Williams, J.M.J. *Tetrahedron Lett.*, 1996, **37**, 1859.

162. For example see *Biomimetic Oxidations Catalysed by Transition Metal Complexes*, Meunier, B. World Scientific Publishing, Imperial College, London, 2000.

Part II: Procedures

2 General Information

Before conducting an asymmetric synthesis, one needs to ensure that one is able to separate both enantiomers of the desired compound. The racemate (*RS*, 50% of each enantiomer) needs to be synthesized in order to study the different possibilities of differentiating each enantiomer.

Two enantiomers can be differentiated by their retention time (R_f) during chromatography on a chiral support, for example, using High Pressure Liquid Chromatography (HPLC) or Gas Chromatography (GC) over a chiral column. They can also be separated by derivatization with an homochiral auxiliary, affording the corresponding diastereomers. As the two diastereomers can have different chemical shifts in NMR spectra, their analysis by ^{1}H-, ^{13}C- or ^{19}F-NMR spectroscopy represents a useful method for the determination of the enantiomeric excess. (*R*)-(+)-α-methoxy-α-(trifluoromethyl)phenylacetic acid or its (*S*)-(−)-enantiomer (MTPA, Mosher acid) is used in the following chapters to determine the enantiomeric excess (e.g. of allylic alcohols). Chemical shift reagents such as [europium(III)-tris[3-(heptafluoropropylhydromethylene)-d-camphorate]] (Eu(hfc)$_3$) can also be used to assess the ratio in a mixture of enantiomers. Each method needs to be performed on the racemic compound in order to find the conditions to separate the two enantiomers.

For experiments conducted in Liverpool GC was performed on a Shimadzu GC-14A gas chromatograph using a SE30 capillary column with the injector and detector set to 250 °C; chiral GC was performed with chiral capillary columns (Lipodex® E and C as indicated) with the injector and detector set to 250 °C. HPLC was performed on a Gilson chromatograph equipped with chiral columns Daicel Chiralpack® AD and OD (wavelength 254 nm).

NMR spectra were recorded on Bruker AC200 spectrometers; unless indicated otherwise deuteriated chloroform was used as solvent and tetramethylsilane as internal reference. Chemical shifts (δ) are given in ppm. The following abbreviations were used to define the multiplicities; s, singlet; d, doublet; t, triplet; q, quartet; m, multiplet; br, broad; coupling constants (*J*) are measured in Hertz (Hz). IR spectra were recorded on a Nicolet Magna-550 FTIR spectrometer. High Resolution Mass Spectra were recorded on a Kratos profile HV3, CIPOS, Fisons VG7070E spectrometer.

Some of the procedures described in the following chapters had to be carried out under an inert atmosphere, nitrogen or argon, to minimize contact with oxygen and moisture. It is then necessary to use Schlenk techniques including the utilization of a vacuum line connected to a high vacuum pump and an inert gas inlet. The use of such equipment requires experience in working under anhydrous conditions.

All the procedures described were performed using dry solvents which were freshly distilled under nitrogen. Tetrahydrofuran and ether were distilled from sodium benzophenone ketal under nitrogen, and dichloromethane from calcium hydride under nitrogen. Petroleum ether (b.p. 40–60 °C) was distilled. Starting materials and solvents were used as obtained from commercial suppliers without further purification unless specified otherwise.

Molecular sieves or magnesium sulfate were activated by heating at 500 °C for 2–14 hours and cooled in a desiccator under vacuum.

Flash column chromatography was performed using Merck 60-silica gel (40–63 μm) and solvents were obtained commercially and used as received.

Most of the reactions described in the following chapters were monitored by Thin Layer Chromatography (TLC) using plastic TLC plates coated with a thin layer of Merck 60 F_{254} silica gel. The products were detected by using an ultraviolet lamp or the TLC plates were treated with p-anisaldehyde reagent, prepared as explained below, and then heated to 120 °C to stain the spots. After visualization and measurement, the R_f values were recorded.

PREPARATION OF p-ANISALDEHYDE REAGENT

Materials and equipment

- Ethanol (370 mL)
- Concentrated sulfuric acid (14 mL)
- Glacial acetic acid (4 mL)
- p-Anisaldehyde

- Beaker, 500 mL with a magnetic stirrer bar
- Magnetic stirrer
- Glass bottle for storage, 500 mL

Procedure

The breaker was filled with ethanol (370 mL); concentrated sulfuric acid (14 mL) was added slowly followed by glacial acetic acid (4 mL) and then p-anisaldehyde (1 mL). The solution was stirred for 15 minutes and then transferred to a suitable labelled bottle for storage.

3 Asymmetric Epoxidation

CONTENTS

3.1 INTRODUCTION

The stereoselective oxidation of organic compounds is dominated by studies of epoxidation (Figure 3.1). Epoxides are useful in organic synthesis, they are versatile intermediates and easily undergo stereoselective ring-opening reactions to form bifunctional compounds[1]. This explains the development of many methods for the synthesis of the enantiomerically pure epoxides. We will describe in Chapters 4–6 different methods of asymmetric epoxidation of functionalized and unfunctionalized alkenes. This chapter expands the information given in Chapter 1 and forms a consolidated introductory section for the next three chapters

In 1980 a useful level of asymmetric induction in the epoxidation of some alkenes was reported by Katsuki and Sharpless[2]. The combination of titanium (IV) alkoxide, an enantiomerically pure tartrate ester and *tert*-butyl hydroperoxide was used to epoxidize a wide variety of allylic alcohols in good yield and enantiomeric excess (usually >90%). This reaction is now one of the most widely applied reactions in asymmetric synthesis[3].

Concerning the oxidation of electron-deficient alkenes such as chalcone derivatives, in 1980, Juliá *et al.* reported an example of a highly stereoselective epoxidation of an electron deficient alkene using a triphasic catalysis system. This method involves alkaline aqueous hydrogen peroxide, an organic solvent and an insoluble polyamino acid.[4–7] A refined method now employs a biphasic system consisting of an oxidant, a non-nucleophilic base, a polyamino acid and an organic solvent[8].

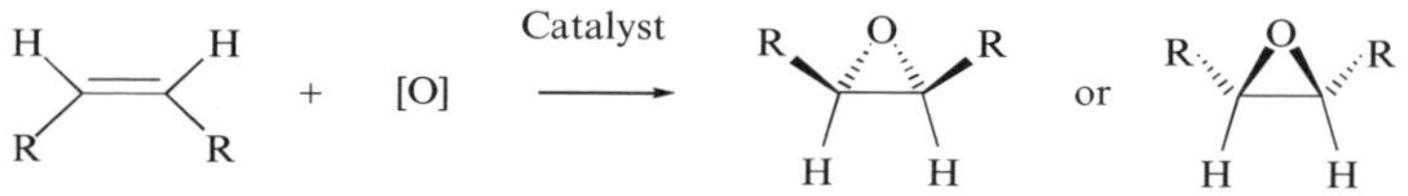

Figure 3.1   Catalytic asymmetric epoxidation of alkenes.

A new method of asymmetric epoxidation of α, β-unsaturated ketones using a stoichiometric amount of N-methylpseudoephedrine as a chiral source in the presence of diethylzinc and oxygen to afford the α, β-epoxy-ketones with good yield and enantiomeric excess was developed by Enders and co-workers[9]. Shibasaki[10] reported an efficient catalytic asymmetric epoxidation of enones using lanthanoid complexes, which give epoxides with enantiomeric excesses between 83 and 94%. This last method will be reported in another volume of this series.

New methods for asymmetric epoxidation of alkenes, bearing no functionality to precoordinate the catalyst, have also been developed successfully in the past few years[11]. Among these methods, Jacobsen et al.[12] were able to epoxidize monosubstituted, disubstituted Z- and trisubstituted alkenes with good asymmetric induction, using cationic (salen)manganese(III) complexes. Shi et al.[13] reported a method of epoxidation using dioxirane generated in situ from potassium peroxomonosulfate and a chiral fructose-derived ketone as catalyst. Using this method high enantioselectivity can be obtained for the epoxidation of unfunctionalized E-alkenes.

Other methods of epoxidation were described; for example in 1979, Groves et al.[14] reported the first example of alkene epoxidation by a chloroferritetraphenylporphyrin catalyst. By adding an optically active group on to this catalyst, they obtained optically active chiral epoxides but generally with a low enantiomeric excess[14]. A number of metalloporphyrins have been used for the epoxidation of unfunctionalized alkenes[15] (see Chapter 6.3). Asymmetric epoxides can also be obtained using enzymes. Peroxidases[16, 17] and monooxygenases[18–20] catalyse the synthesis of nonracemic chiral epoxides. A kinetic resolution of racemic epoxides can be catalysed by epoxide hydrolases.[21–23] Those methods (using enzymes) will not be described in this chapter since enzymatic epoxidation has been reviewed previously[24].

Chapters 4–6 present an overview and a comparison between the various existing strategies for asymmetric epoxidation of unfunctionalized alkenes, α, β-unsaturated ketones and allylic alcohols.

REFERENCES

1. Besse, P., Veschambre, H. *Tetrahedron*, 1994, **50**, 8885.
2. Katsuki, T., Sharpless, K.B. *J. Am. Chem. Soc.*, 1980, **102**, 5974.
3. Johnson, R.A., Sharpless, K.B. *Catalytic Asymmetric Epoxidation of Allylic Alcohols*; VCH, Ed.; Ojima, 1993, pp S 103.
4. Juliá, S., Masana, J., Vegas, J.C. *Angew. Chem. Int. Ed. English*, 1980, **11**, 929.
5. Colonna, S., Molinari, H., Banfi, S., Juliá, S., Mansana, J., Alvarez, A. *Tetrahedron*, 1983, **39**, 1635.
6. Juliá, S., Guixer, J., Mansana, J., Rocas, J., Colonna, S., Annuziata, R., Molonari, H. *J. Chem. Soc., Perkin Trans. I*, 1982, 1317.

7. Banfi, S., Colonna, H., Molinari, R., Juliá, S., Guixer, J. *Tetrahedron*, 1984, **40**, 5207.
8. Bentley, P.A., Bergeron, S., Cappi, M.W., Hibbs, D.E., Hursthouse, M.B., Nugent, T.C., Pulido, R., Roberts, S.M., Wu, L.E. *J. Chem. Soc., Chem. Commun.*, 1997, 739.
9. Enders, D., Zhu, J., Kramps, L. *Liebigs Ann./ Recueil* 1997, 1101.
10. Bougauchi, M., Watanabe, S., Arai, T., Sasai, H., Shibasaki, M. *J. Am. Chem. Soc.*, 1997, **119**, 2329.
11. Bolm, C. *Angew. Chem. Int. Ed. English* 1991, **30**, 403.
12. Jacobsen, E.N., Zhang, W., Muci, A.R., Ecker, J.R., Deng, L. *J. Am. Chem. Soc.*, 1991, **113**, 7063.
13. Wang, Z.-X., Tu, Y., Frohn, M., Zhang, J.-R., Shi, Y. *J. Am. Chem. Soc.*, 1997, **119**, 11224.
14. Groves, J.T., Nemo, T.E., Myers, R.S. *J. Am. Chem. Soc.*, 1979, **101**, 1032.
15. Meunier, B. *Chem. Rev.* 1992, **92**, 1411.
16. Allain, E.J., Lowell, P.H., Hager, L.P., Deng, L., Jacobsen, E.N. *J. Am. Chem. Soc.*, 1993, **115**, 4415.
17. Dexter, A.F., Lakner, F.J., Campbell, R.A., Hager, L.P. *J. Am. Chem. Soc.*, 1995, **117**, 6412.
18. Colbert, J.E., Katopodis, A.G., May, S.W. *J. Am. Chem. Soc.*, 1990, **112**, 3993.
19. De Smet, M.-J., Witholt, B., Wynberg, H. *J. Org. Chem.*, 1981, **46**, 3128.
20. Fu, H., Newcomb, M., Wong, C.-H. *J. Am. Chem. Soc.*, 1991, **113**, 5878.
21. Mischitz, M., Kroutil, W., Wandel, U., Faber, K. *Tetrahedron: Asymmetry*, 1995, **6**, 1261.
22. Pedragosa-Moreau, S., Archelas, A., Furstoss, R. *J. Org. Chem.*, 1993, **58**, 5533.
23. Wandel, U., Mischitz, M., Kroutil, W., Faber, K. 1995, *J. Chem. Soc., Perkin Trans. I*, 735.
24. Roberts, S.M. *Preparative Biotransformations* Wiley, Chichester 1997.

4   Epoxidation of α, β-Unsaturated Carbonyl Compounds

CONTENTS

4.1   NON-ASYMMETRIC EPOXIDATION

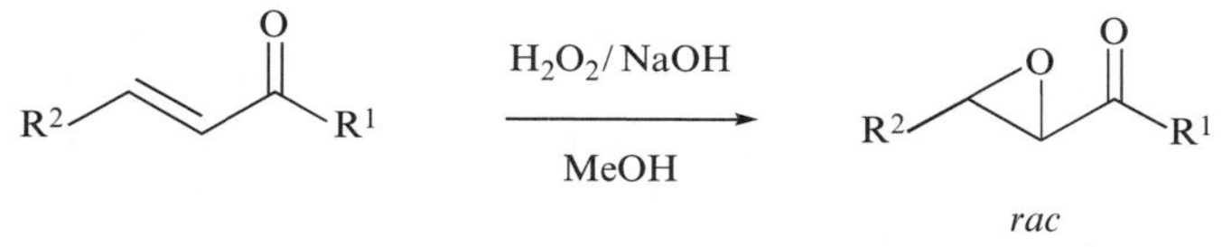

As α, β-unsaturated ketones are electron-poor alkenes, they do not generally give epoxides when treated with peracids. They can be epoxidized with hydrogen peroxide which involves nucleophilic attack by HOO⁻ to give the epoxy ketone (Figure 4.1).

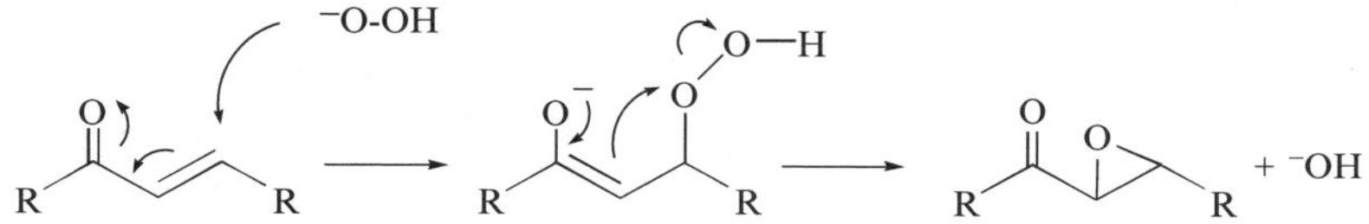

Figure 4.1   Mechanism of α, β-unsaturated ketone epoxidation.

Materials and equipment

- α, β-Unsaturated ketone, 2 mmol
- Sodium hydroxide, 100 mg, 2.5 mmol, 1.25 eq
- Anhydrous methanol, 10 mL
- Solution of 30 % of hydrogen peroxide, 300 mg, 2.5 mmol, 1.25 eq

  Hydrogen peroxide can cause burns: wear suitable protective clothing, including eye and face protection. Store in a cool place.

- Brine
- Dichloromethane
- Magnesium sulfate
- Silica gel 60 (0.063–0.04 mm)

- 50 mL Round-bottomed flask with a magnetic stirrer bar
- Magnetic stirrer
- Separating funnel, 250 mL
- Rotary evaporator

Procedure

1. In a 50 mL dry round-bottomed flask was dissolved the α, β-unsaturated ketone (2 mmol) in anhydrous methanol (10 mL); hydrogen peroxide (300 mg) was added.
2. The reaction mixture was stirred at room temperature and the reaction monitored by TLC. After completion, the reaction was carefully quenched with water (10 mL). A white precipitate appeared.
3. The reaction mixture was transferred into a separating funnel and the aqueous layer extracted with dichloromethane (3 × 30 mL). The combined organic layers were washed with water (3 × 30 mL) and then with brine (30 mL), dried over magnesium sulfate, filtered and concentrated under reduced pressure.
4. The residue was purified by flash chromatography on silica gel as required. (See below for the purification methods for each substrate.)

4.2 ASYMMETRIC EPOXIDATION USING POLY-D-LEUCINE

Bentley et al.[1] recently improved upon Juliá's epoxidation reaction. By using urea–hydrogen peroxide complex as the oxidant, 1,8-diazabicyclo[5,4,0]undec-7-ene (DBU) as the base and the Itsuno's immobilized poly-D-leucine (Figure 4.2) as the catalyst, the epoxidation of α, β-unsaturated ketones was carried out in tetrahydrofuran solution. This process greatly reduces the time required when compared to the original reaction using the triphasic conditions.

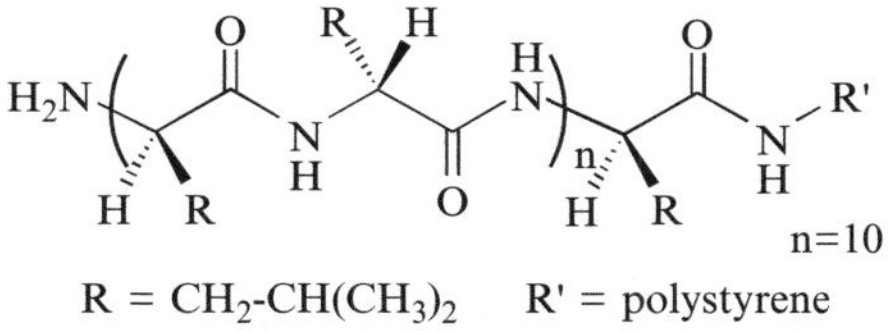

R = CH$_2$-CH(CH$_3$)$_2$ R' = polystyrene

Figure 4.2 Immobilized poly-D-leucine catalyst.

4.2.1   SYNTHESIS OF LEUCINE N-CARBOXYANHYDRIDE

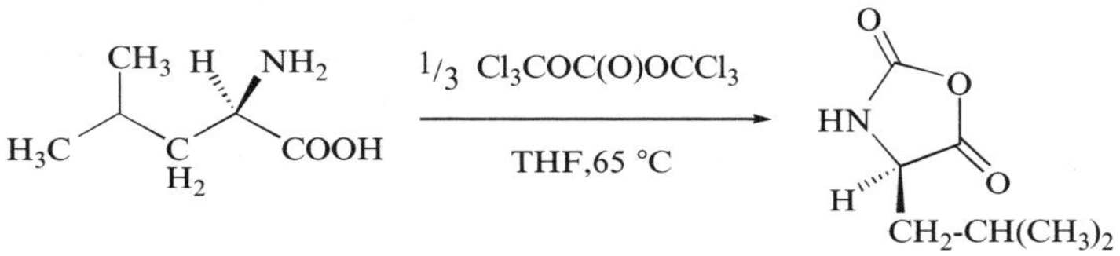

Materials and equipment

- D-Leucine, 7.00 g, 53.4 mmol
- Dry tetrahydrofuran, 115 mL
- Triphosgene, 6.34 g, 21.4 mmol
- n-Hexane (800 mL), diethyl ether (100 mL)
- Celite®
- Silica gel 60 (0.063–0.04 mm)
- Sand

- 250 mL Two-necked and 500 mL round-bottomed flask with magnetic stirrer bars
- Magnetic stirrer
- Reflux condenser
- Magnetic stirrer with thermostatically controlled oil bath and thermometer
- Glass sintered funnel, diameter 5.5 cm
- Rotary evaporator

Procedure

1. The 250 mL round-bottomed flask, equipped with a magnetic stirrer bar, was dried in an oven at 120 °C for 4 hours. The flask was removed, sealed, cooled under vacuum and flushed with nitrogen.
2. D-Leucine (7.00 g) was placed in the flask and vacuum was applied to the system. The flask was flushed with nitrogen for 1 hour and equipped with a reflux condenser. Tetrahydrofuran (85 mL) was added and the mixture heated to a gentle reflux (b.p. 65–67 °C).

- *The solution remained cloudy due to the insolubility of D-leucine even when heated in tetrahydrofuran.*
3. When the mixture was refluxing, the triphosgene (5.81 g) was added carefully, portionwise over 5 minutes. The mixture was heated for 1 hour.

   Triphosgene is very toxic. Wear suitable gloves, and eye and face protection. Handle very carefully in a well-ventilated fume-hood.

   The mixture gradually became clearer as the insoluble material was consumed. If the mixture is not clear after 1 hour continue heating for another 20 minutes. If the mixture at this stage remained unclear, 0.53 g of triphosgene was added and heated further.
4. Once the clear mixture had cooled to room temperature, it was then poured into a 500 mL round-bottomed flask containing *n*-hexane (400 mL).
5. A sintered glass funnel was packed with 2 cm Celite®, 1 cm silica gel and 3 cm sand. The *n*-hexane solution was carefully filtered through this packed funnel without disturbing the packing material. The filtration was completed by rinsing the packing with diethyl ether (100 mL).
6. The solvent was removed from the filtrate using a rotary evaporator to afford a white solid. Dry tetrahydrofuran (30 mL) was added to dissolve the solid and then *n*-hexane was added until a white solid precipitated.

   Approximately 400 mL of *n*-hexane was necessary to precipitate the product.
7. The solution was filtered and the resulting white solid was dried under high vacuum. This provided leucine *N*-carboxyanhydride (6.7 g, 80%); m.p. 76–77 °C, (Lit.[2] 77–79 °C).

   This procedure had been scaled up to provide 50 g of the *N*-carboxyanhydride.

4.2.2   SYNTHESIS OF IMMOBILIZED POLY-D-LEUCINE

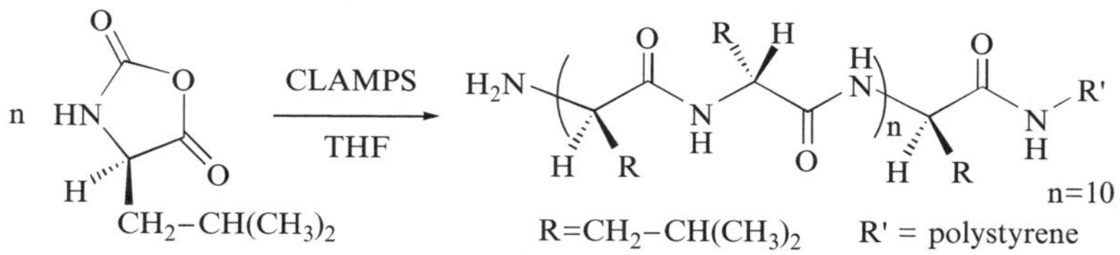

Materials and equipment

- Tetrahydrofuran, 110 mL
- *N*-Carboxyanhydride
- *Cross*-linked aminomethylpolystyrene (CLAMPS), 500 mg[3, 4]
- Acetone–distilled water (1:1), 50 mL

- Acetone–distilled water (4:1), 50 mL
- Diethyl ether (50 mL), acetone (100 mL), ethyl acetate (50 mL), petroleum ether (50 mL)

- 250 mL Round-bottomed flask with a magnetic stirrer bar
- Magnetic stirrer

Procedure

1. The *N*-carboxyanhydride was placed in a 250 mL round-bottomed flask equipped with a magnetic stirrer bar, under nitrogen. Dry tetrahydrofuran (110 mL) was then added followed by *cross*-linked aminomethylpolystyrene. The mixture was stirred for 4 days.
2. The solution was filtered, diethyl ether (50 mL) was added to the solid and collected on the filter. The solid was left to soak in the diethyl ether for 30 minutes before removing the solvent by aid of a water aspirator. This procedure of soaking the solid on the filter was repeated with acetone–distilled water (1:1), acetone–distilled water (4:1), acetone (2 × 50 mL), ethyl acetate (50 mL) and petroleum ether (50 mL).
3. The white solid was placed under high vacuum overnight to give immobilized poly-D-leucine (4.4 g) as a white powder.

   The quality of the catalyst can be determined by performing an asymmetric epoxidation reaction on chalcone according to the following procedure. The activity of the polymer was considered satisfactory if it provided the epoxy-chalcone in 85% yield and 95% of enantiomeric excess, with a reaction time between 10 and 40 minutes.

4.2.3 ASYMMETRIC EPOXIDATION OF (*E*)-BENZYLIDENEACETOPHENONE

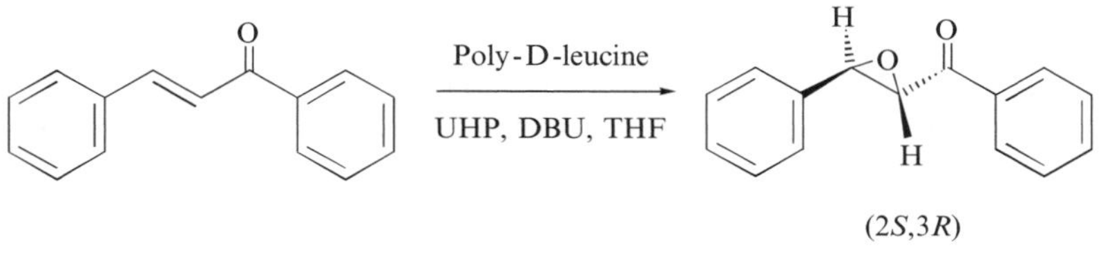

(2*S*,3*R*)

Materials and equipment

- Anhydrous tetrahydrofuran, 0.8 mL
- Chalcone (*E*)-benzylideneacetophenone (97%), 50 mg, 0.24 mmol
- Poly-D-leucine, 100 mg
- Urea–hydrogen peroxide (UHP, 98%), 27 mg, 0.28 mmol

- 1,8-Diazabicyclo[5.4.0]undec-7-ene (DBU, 98%) three drops, 90 mg, 32 mmol
- Ethyl acetate, petroleum ether
- Brine
- Magnesium sulfate

- Two 10 mL round-bottomed flasks with magnetic stirrer bars
- Magnetic stirrer
- Büchner funnel (5 cm)
- Büchner flask
- Filter paper
- Separating funnel, 250 mL
- Rotary evaporator

Procedure

1. In a 10 mL round-bottomed flask equipped with a magnetic stirrer bar were placed tetrahydrofuran (0.8 mL) and immobilized poly-D-leucine (100 mg). (E)-Benzylideneacetophenone (50 mg), 1,8-diazabicyclo[5.4.0] undec-7-ene (90 mg), and urea–hydrogen peroxide (27 mg) were added to the mixture. The thick white reaction mixture was stirred vigorously for 30 minutes.
2. The reaction was monitored by TLC (eluent: petroleum ether–ethyl acetate, 4:1). Visualized by p-anisaldehyde dip, the chalcone stained brown, R_f 0.42 and the epoxide stained blue, R_f 0.36.
3. After completion, the mixture was poured into ethyl acetate and the catalyst was removed by filtration using two filter papers in a Büchner funnel. The poly-D-leucine residue was washed with ethyl acetate (2 × 10 mL), with water (2 × 10 mL) and with brine (10 mL).
4. The mixture was transferred into a separating funnel and the phases were separated. The aqueous layer was extracted with ethyl acetate (3 × 20 mL). The combined organic layers were dried over magnesium sulfate, filtered and concentrated using a rotary evaporator to give a crystalline solid (2S,3R)-epoxychalcone (90%).

   The ee (95–99%) was determined by HPLC (Chiralpak® AD column, flow 1 mL/min, eluent ethanol–n-hexane 1:9); (2R,3S)-enantiomer: R_t 13.6 min, (2S,3R)-enantiomer: R_t 20.5 min.

   ^{1}H NMR (200 MHz, CDCl$_3$): δ 7.3–7.6, 7.9–8.0 (2m, 10H, Ph); 4.24 (d, J 2.0 Hz, 1H, H$_\alpha$); 4.02 (d, J 2.0 Hz, 1H, H$_\beta$).

   IR (CHCl$_3$, cm^{-1}): 3069 (C–H epoxide), 3012, (C–H aromatic), 2965, 1692 (C=O), 1599, 1581, 1450 (C=C), 1406 (C–C aromatic), 1260, 1203 (C–O–C), 1098, 1009, 884.

4.2.4   CONCLUSION

The validation was performed without any modifications. The procedure is very easy to reproduce and the achieved results correlate with the published material. The yield is somewhat lower than the published result, though monitoring of the reaction by TLC indicated complete consumption of substrate. This is believed to be due to decreased precipitation during recrystallization. Because the product is unstable in solution it is recommended that the recrystallization is performed as quickly as possible. Alternatively, the impurities can be removed by trituration.

The asymmetric epoxidation reaction with polyleucine as catalyst may be applied to a wide range of α, β-unsaturated ketones. Table 4.1 shows different chalcone derivatives that can be epoxidized with poly-L-leucine. The substrate range included dienes and tetraenes[5]. Some other examples were reported in a previous edition[6] and by M. Lasterra-Sánchez[7].

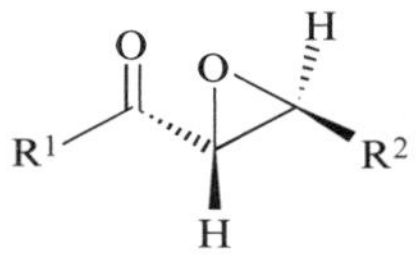

4.3   ASYMMETRIC EPOXIDATION USING CHIRAL MODIFIED DIETHYLZINC[8]

The epoxidation of enones with poly-D-leucine is complementary to other strategies. Enders et al.[8] introduced a new method for the asymmetric epoxidation of α-enones using diethylzinc, oxygen and (1R, 1R)-or (1S, 2S)-N-methylpseudoephedrine as chiral auxiliary.

The mechanism of this 'one-pot reaction' is proposed to be as follows (Figure 4.3); firstly, a chiral alkoxide ethylzinc is prepared from diethylzinc and the chiral alcohol with the evolution of a gas, which is probably ethane (I). The chiral ethylperoxyzinc alkoxide is formed by insertion of oxygen into the carbon–zinc

Table 4.1   Epoxidation of enones using poly-L-leucine catalyst[5,7].

R^1	R^2	Yield%	ee%
t-Bu	Ph	92	>98 (2R,3S)
i-Pr	Ph	60	62 (2R,3S)
Cyclopropyl	Ph	85	77 (2R,3S)
2-Naphthyl	Cyclopropyl	73	>98 (2R,3S)
2-Naphthyl	Ph–CH=CH	78	>96 (2R,3S)

bond (**II**). This species represents an asymmetric epoxidizing reagent for α, β-unsaturated ketones.

$$Et_2Zn + R^*OH \longrightarrow EtZnOR^* + EtH \qquad (I)$$

$$EtZnOR^* + O_2 \longrightarrow EtOOZnOR^* \qquad (II)$$

R*OH(1*S*,2*S*):

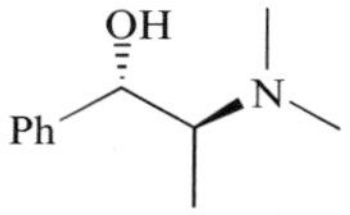

Figure 4.3 Formation of the active species in the epoxidation reaction using diethylzinc and (1*S*, 2*S*)-methylpseudoephedrine.

In the epoxidation process (Figure 4.4), the oxygen of the enone's carbonyl function first coordinates with the zinc atom. The ethylperoxy anion then attacks the β-position, which constitutes a Michael-type addition. The subsequent cyclization gives the epoxy ketone and the zinc alkoxide.

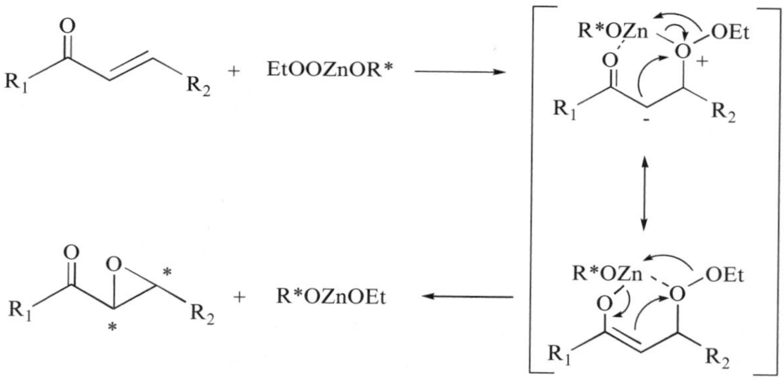

Figure 4.4 Mechanism of epoxidation using diethylzinc and methylpseudoephedrine.

4.3.1 EPOXIDATION OF 2-ISOBUTYLIDENE-1-TETRALONE

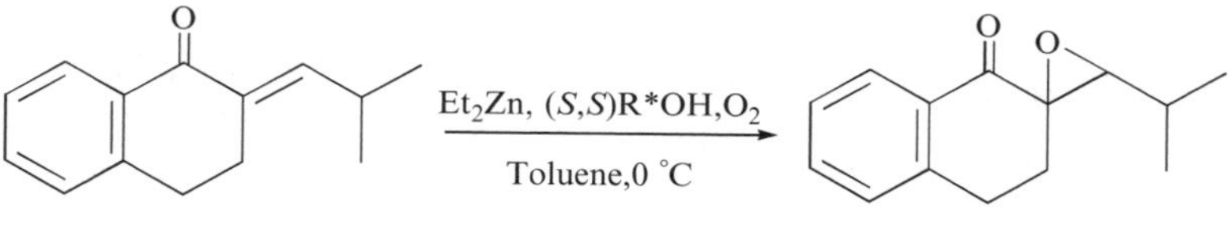

(2*S*, 3*R*)

Materials and equipment

- (1S,2S)-N-Methylpseudoephedrine, 430 mg, 2.4 mmol, 2.4 eq
- Anhydrous toluene, 12 mL
- Diethylzinc solution, 1.1 M in toluene, 1 mL, 1.1 mmol, 1.1 eq

    Diethylzinc is flammable, reacts with water and can cause severe burns. Wear gloves and eye protection, and handle with care.

    If low yield and/or enantiomeric excess is obtained, the purity of the diethylzinc should be the primary suspect.
- Oxygen gas
- 2-Isobutylidene-1-tetralone, 200 mg, 1 mmol

    This substrate was prepared by aldol condensation of tetralone with *iso*-butyraldehyde in the presence of aqueous sodium hydroxide[9].
- Aqueous phosphate buffer solution pH 7, 8 mL
- Dichloromethane, petroleum ether, diethyl ether, *n*-hexane
- Brine
- Sodium sulfate
- Silica gel 60 (0.063–0.04 mm)
- *p*-Anisaldehyde

- 50 mL Two-necked dry flask with a magnetic stirrer bar
- Magnetic stirrer
- Acetone cooling bath equipped with a contact thermometer, 0 °C
- Syringes
- Balloon equipped with a syringe
- Ice-bath
- Separating funnel, 250 mL
- Rotary evaporator

Procedure

1. A 50 mL two-necked flask, equipped with a magnetic stirrer bar, was placed in an oven at 120 °C for 4 hours. The flask was removed, cooled under vacuum and flushed with nitrogen.
2. In the dry flask was dissolved (1S, 2 S)-N-methylpseudoephedrine (430 mg) in anhydrous toluene (10 mL) under argon. The reaction mixture was placed in a cooling bath equipped with a contact thermometer stirred and cooled to 0 °C. Diethylzinc (1 mL) was added carefully *via* a syringe (**evolution of ethane**).
3. After 80 minutes the connection with argon was replaced by a balloon filled with oxygen. The reaction mixture was stirred for 2.5 hours without removing the balloon filled with oxygen.
4. The bath was cooled to −78 °C and a solution of 2-*iso*butylidene-1-tetralone (200 mg) in anhydrous toluene (2 mL) was added via a syringe to the cold mixture. The reaction mixture was stirred at this temperature for 30 minutes

and then rapidly warmed to 0 °C by means of an ice-bath while stirring was continued.

5. The reaction was monitored by TLC (eluent: *n*-hexane–diethyl ether, 8:2). 2-Isobutylidene-1-tetralone (UV active) stained light purple with *p*-anisaldehyde dip, R_f 0.66 and the epoxide (UV active) dark purple, R_f 0.30.

6. The reaction was quenched after 16 hours by addition of aqueous phosphate buffer solution (pH 7, 8 mL).

   It is recommended to replace the oxygen balloon for an argon balloon after 3.5 hours; extensive exposure to O_2 can have an adverse influence on the reaction.

7. The reaction mixture was transferred into a separating funnel and the two layers were separated. The upper aqueous layer was extracted with dichloromethane (3 × 30 mL). The combined organic layers were washed with brine, dried over sodium sulfate and the solvent was evaporated under reduced pressure.

8. The residue was purified by flash chromatography using petroleum ether–diethyl ether (9:1) as eluent to give (2S, 3′R)-1,2,3,4-tetrahydro-3′-*iso*propylspiro [naphthalene-2,2′-oxirane]-1-one as a yellow oil (190 mg, 0.88 mmol, 90 %).

   The ee (>99 %) was determined by HPLC (Chiralpak® AD column, flow 1 mL/min, ethanol–*n*-hexane; 1:9); (2S, 3′R)-enantiomer: R_t 8.84 min, (2R, 3′S)-enantiomer: R_t 7.02 min.

   ¹H NMR (200 MHz, CDCl₃): δ 8.09 (dd, J 7.7 Hz, J 1.1 Hz, 1H, COCCH); 7.54 (td, J 7.4 Hz, J 1.6 Hz, 1H, COCCHCH); 7.34 (m, 1H, CH₂CCHCH); 3.15 (dd, J 8.2 Hz, J 4.4 Hz, 2H, COCCH₂CH_2); 3.00 (d, J 9.3 Hz, 1H, CH₃CHCH); 2.50 (dt, J 13.7 Hz, J 8.4 Hz, 1H, COCCH_2); 2.14 (dt, J 13.7 Hz, J 4.7 Hz, 1H, COCCH_2); 1.68 (m, 1H, CH(CH₃)₂); 1.2 (d, J 6.6 Hz, 3H, CH₃); 2.06 (d, J 6.6 Hz, 3H, CH₃).

   IR (CHCl₃, cm⁻¹): 3009 (C–H aromatic), 2972, 2935, 2873 (C–H aliphatic), 1687 (C = O), 1604 (C–C aromatic), 1469, 1457, 1433 (CH₂, CH₃), 1316, 1304, 1203 (C–O–C), 1158, 925, 893, 878, 844, 822.

   Mass: calculated for $C_{14}H_{16}O_2$: *m/z* 216.11502; found [M]⁺• 216.11520.

4.3.2   CONCLUSION

Ender's method is easy to reproduce; however, it needs a freshly prepared diethylzinc solution as its quality can dramatically influence the enantiomeric excess. Strictly, the reaction is not a catalytic process but the methylpseudoephedrine (chiral auxiliary) can be recovered almost completely with unchanged enantiomeric purity during the flash chromatography. This method gives good results for epoxidation of α-enones such as (E)-2-alkyliden-1-oxo-1,2,3,4-tetrahydronaphthalene. The enantiomeric excess obtained during the validation correlates with the published result. Table 4.2 gives the results obtained by the method described above, according to the relevant publication.

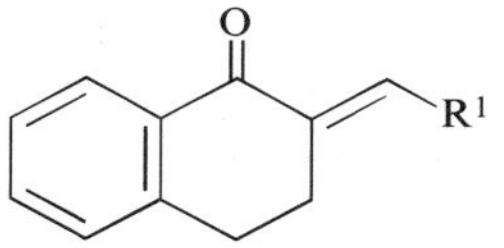

Table 4.2 α,β-Epoxy ketones prepared by epoxidation of (E)-2-alkyliden-1-oxo-1,2,3,4-tetrahydronaphthalenes using diethylzinc and (1R,2R)-N-methylpseudoephedrine[8].

R^1	Yield %	ee %
H	40	3
Me	85	80
Et	65	90
i-Pr	98*	>99*
n-Pr	99	83
Ph	62	64

*Reaction validated

For epoxidation of chalcones using Ender's method, the results depend on the nature of the substrate. For the (E)-benzylideneacetophenone (R^1, R^2 = Ph), the enantiomeric excess was only 60% using the same procedure as the one described above, whereas the polyleucine method furnished the epoxide with an enantiomeric excess > 95%. Table 4.3 gives some results of the epoxidation of some acyclic enones using Ender's method.

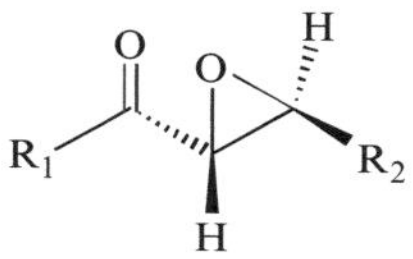

Table 4.3 α, β-Epoxy ketones prepared by epoxidation of α-enones using diethylzinc and (1R, 2R)-N-methylpseudoephedrine (according to the relevant publication)[8].

R^1	R^2	Yield %	ee % (configuration)
Ph	Et	99	91 (2R,3S)
Ph	Me	96	85 (2R,3S)
Ph	n-Pr	99	87 (2R,3S)
Ph	i-Pr	97	92 (2R,3S)
Ph	Ph	94*	61 (2R,3S)*
t-Bu	Ph(CH$_2$)$_2$	99	90 (2R,3S)
2,4,6-Me$_3$Ph	Et	94	82 (2R,3S)

* Reaction validated

4.4 ASYMMETRIC EPOXIDATION OF (E)-BENZYLIDENEACETOPHENONE USING THE La-(R)-BINOL-Ph₃PO/CUMENE HYDROPEROXIDE SYSTEM

K. DAIKAI, M. KAMAURA, and J. INANAGA

Institute for Fundamental Research of Organic Chemistry (IPOC), Kyushu University Hakozaki, Fukuoka 812–8581, Japan, e-mail: inanaga@ms.ifoc.kyushu-u.ac.jp

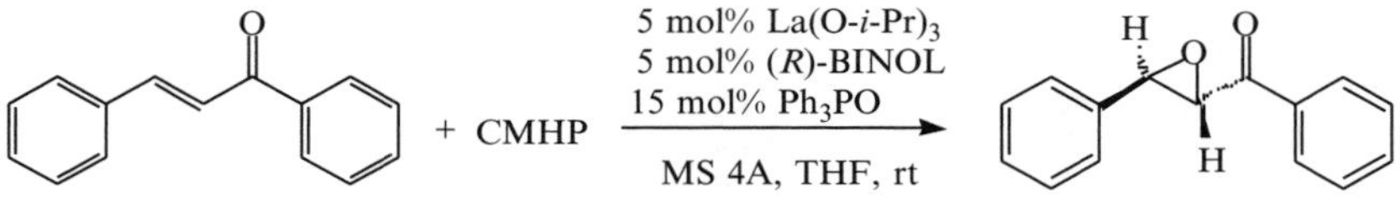

(2S, 3R)

Materials and equipment

- Anhydrous tetrahydrofuran, 3.2 mL
- Chalcone, (E)-benzylideneacetophenone (97%), 93.5 mg, 0.449 mmol
- (R)-(+)-1,1'-Bi-2-naphthol [(R)-BINOL, 99%], 6.4 mg, 0.0224 mmol
- Triphenylphosphine oxide (Ph₃PO, 99%), 18.7 mg, 0.0672 mmol
- Tri*iso*propoxylanthanum (La(O–i-Pr)₃, 99%), 7.1 mg, 0.0225 mmol
- Cumene hydroperoxide (CMHP, 99%), 99.6 μL, 0.673 mmol
- Molecular sieves 4 Å, 44.9 mg
- Ethyl acetate, hexane
- Silica gel
- Celite

- 10 mL Round-bottomed flask with a magnetic stirrer bar
- Magnetic stirrer
- Syringe
- Filter tube
- Rotary evaporator
- Glass column

Procedure

1. In a 10 mL round-bottomed flask equipped with a magnetic stirrer bar were placed activated molecular sieves 4A (44.9 mg), (R)-BINOL (6.4 mg), triphenylphosphine oxide (18.7 mg), and anhydrous tetrahydrofuran (0.9 mL) and the mixture was stirred for 5 minutes under argon. To this suspension was added a suspension of La(O–i-Pr)₃ (7.1 mg) in tetrahydrofuran (1.4 mL) by a syringe. After stirring for 1 hour at room temperature, cumene hydro-

peroxide[10] (99.6 μL) was added and the whole mixture was stirred for 30 minutes. To the resulting faintly green suspension was added a tetrahydrofuran (0.9 mL) solution of (E)-benzylideneacetophenone (93.5 mg) and the mixture was stirred vigorously for 15 minutes at room temperature.

2. The reaction was monitored by TLC (SiO$_2$, eluent: hexane–ethyl acetate, 4:1), where both the substrate and the product were detected by a UV lamp and also visualized by 10% ethanolic phosphomolybdic acid dip: chalcone, R_f 0.46; the epoxide, R_f 0.37.

3. After completion of the reaction, silica gel (ca. 100 mg) was added while stirring. The reaction mixture was filtered through Celite (ca. 300 mg) and washed with ethyl acetate (ca. 5 mL).

4. The filtrate was concentrated using a rotary evaporator and the residue was purified by column chromatography on silica gel (eluent: hexane–ethyl acetate, 30:1) to give (2S, 3R)-epoxychalcone (99%) as a colourless solid.

The ee (99.7%) of epoxychalcone was determined by HPLC (DAICEL CHIRALCEL OB-H, eluent 2-propanol–hexane 1:2, flow rate 0.5 mL/min); (2S, 3R)-enantiomer: R_t 23.4 min, (2R, 3S)-enantiomer: R_t 30.6 min (detection at 254 nm with D$_2$ lamp).

^{1}H NMR (400 MHz, CDCl$_3$): δ 8.03–8.01 (m, 2H); 7.65–7.61 (m, 1H); 7.52–7.28 (m, 7H); 4.31 (d, J 1.96 Hz, 1H, H$_\alpha$); 4.09 (d, J 1.96 Hz, 1H, H$_\beta$).

Notes

The amount of molecular sieves 4 Å largely influences the product's ee[11]. Usually 100 mg (for the CMHP oxidation) or 1 g (for the TBHP oxidation) of MS 4Å for 1 mmol of substrate is enough; however, in the case where chemical yield and/or optical yield are not high, use of excess MS 4Å often improves them. The addition of achiral ligands such as tributylphosphine oxide, tri-o-tolyl- and tri-p-tolylphosphine oxides, hexamethylphosphoric triamide, triphenylphosphate, lutidine N-oxide, and 1,3-dimethyl-2-imidazolidinone were found to be less effective than that of triphenylphosphine oxide in the epoxidation of chalcone.

The method is applicable to a wide range of substrates. Table 4.4 gives various α, β-enones that can be epoxidized with the La-(R)-BINOL-Ph$_3$PO/ROOH system. The substituents (R^1 and R^2) can be either aryl or alkyl. Cumene hydroperoxide can be a superior oxidant for the substrates with R^2 = aryl group whereas t-butyl hydroperoxide (TBHP) gives a better result for the substrates with R^1 = R^2 = alkyl group.

Table 4.4 Epoxidation of α, β-enones using a La-(R)-BINOL-Ph₃PO catalyst system[11,12].

R¹	R²	ROOH	Time/h	Yield/%	Ee/%
Ph	Ph	CMHP	0.25	99	99.7
Ph	Ph	TBHP	0.5	99	96
i-Pr	Ph	CMHP	4	72	>99.9
i-Pr	Ph	TBHP	12	67	96
Me	Ph	CMHP	3	90	99.5
Me	Ph	TBHP	6	92	93
Ph	i-Pr	CMHP	3	88	98
Ph	i-Pr	TBHP	0.5	87	93
Me	Ph(CH₂)₂	CMHP	18	56	85
Me	Ph(CH₂)₂	TBHP	1	92	87

The amount of the catalyst can be reduced to 1 mmol% without reducing the enantioselectivity considerably: 99% ee (98% yield) of epoxychalcone was obtained in the epoxidation of chalcone with CMHP.

As shown in Figure 4.5, a remarkably high positive nonlinear effect was observed in the La-BINOL-Ph₃PO complex-catalysed epoxidation of chalcone (either with CMHP or with TBHP as an oxidant)[12], which strongly suggests that the active catalyst leading to high enantioselection does not have a monomeric structure but may exist as a thermodynamically stable dinuclear complex.

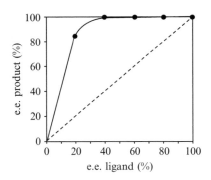

Figure 4.5 Nonlinear effect in the epoxidation of chalcone using the La-(R)-BINOL-Ph₃PO/CMHP system.

4.4.1   MERITS OF THE SYSTEM

● The method has wide applicability and can be carried out under mild conditions.

- All of the reagents required for the asymmetric epoxidation are commercially available.
- Both enantiomers of α, β-epoxy ketones can be synthesized essentially at the same cost since both (R)- and (S)-BINOLs are sold at almost the same price.
- It may not be necessary to employ an optically pure chiral ligand (BINOL) for the preparation of the catalyst because a high degree of asymmetric amplification can be expected.

REFERENCES

1. Bentley, P.A., Bergeron, S., Cappi, M.W., Hibbs, D.E., Hursthouse, M.B., Nugent, T.C., Pulido, R., Roberts, S.M., Wu, L.E. *J. Chem. Soc., Chem. Commun.*, 1997, 739.
2. Daly, W.H., Poché, D. *Tetrahedron Lett.*, 1988, **29**, 5859.
3. Itsuno, S., Sakakura, M., Ito, K. *J. Org. Chem.*, 1990, **55**, 6047.
4. Bentley, P.A., Kroutil, W., Littlechild, J.A., Roberts, S.M. *Chirality*, 1997, **9**, 198.
5. Kroutil, W., Mayon, P., Lasterra-Sánchez, M.E., Maddrell, S.J., Roberts, S.M., Thornton, S.R., Todd, C.J., Tüter, M. *J. Chem. Soc., Chem. Commun.*, 1996, 845.
6. Roberts, S.M. *Preparative Biotransformations*, Wiley, Chichester 1997.
7. Lasterra-Sánchez, M., Roberts, S.M. *Current Organic Chemistry*, 1997, **1**, 187.
8. Enders, D., Zhu, J., Kramps, L. *Liebigs Ann./ Recueil*, 1997, 1101.
9. Kher, S.M., Kulkarni, G.H. *Synth. Commun.*, 1990, **20**, 2573.
10. Pure CMHP was obtained by the method in *Purification of Laboratory Chemicals*, 4th ed.; Perrin, D.D., Armarego, W.L., Eds; Butterworth-Heinemann: Oxford, pp. 154, 1996.
11. Daikai, K., Kamaura, M., and Inanaga, J., *Tetrahedron Lett.*, 1998, **39**, 7321.
12. Daikai, K., and Inanaga, J., to be published.

5 Epoxidation of Allylic Alcohols

CONTENTS

In 1980, Katsuki and Sharpless[1] reported that, with the combination of a titanium(IV) alkoxide, an enantiomerically pure tartrate ester [for example (+)-diethyl tartrate ((+)-DET) or (+) di-iso-propyltartrate ((+)-DIPT)] and *tert*-butyl hydroperoxide, they were able to carry out the epoxidation of a variety of allylic alcohols in good yield and with a good enantiomeric excess (Figure 5.1).

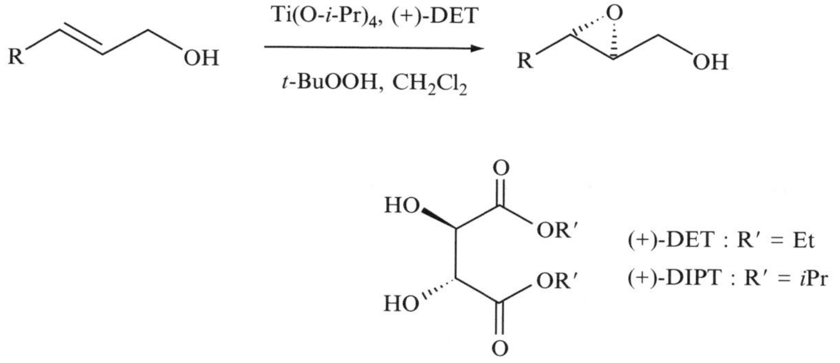

(+)-DET : R' = Et
(+)-DIPT : R' = iPr

Figure 5.1 Allylic alcohol epoxidation using a chiral titanium(IV) complex.

Of fundamental importance to an understanding of the reaction and its mechanism is the fact that in solution there is rapid exchange of titanium ligands (Figure 5.2). After formation of the [titanium$(OR)_2$(tartrate)] complex, the two remaining alkoxide ligands are replaced in reversible exchange reactions by the *tert*-butyl hydroperoxide (TBHP) and the allylic alcohol to give the [titanium(tartrate)(allylic alcohol)(TBHP)] complex. The oxygen is then transferred from the coordinated hydroperoxide to the allylic alcohol[2].

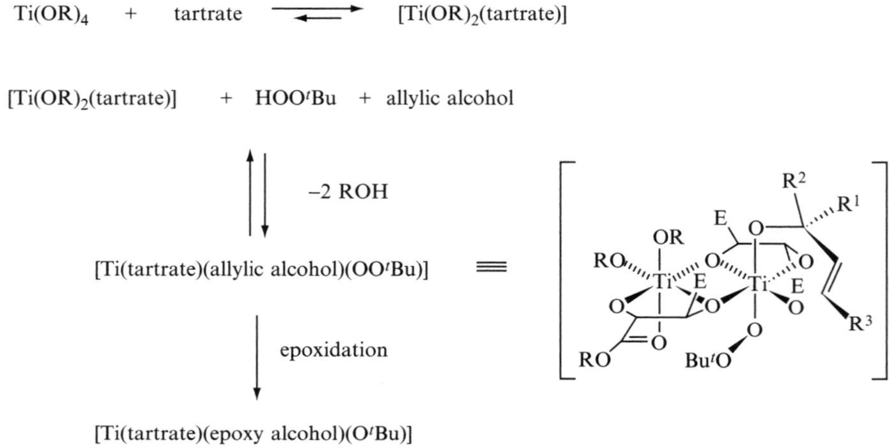

$Ti(OR)_4$ + tartrate $\rightleftharpoons$ [Ti$(OR)_2$(tartrate)]

[Ti$(OR)_2$(tartrate)]   + HOO^{t}Bu + allylic alcohol

$\Big\updownarrow$ -2 ROH

[Ti(tartrate)(allylic alcohol)(OO^{t}Bu)] $\equiv$

$\Big\downarrow$ epoxidation

[Ti(tartrate)(epoxy alcohol)(O^{t}Bu)]

Figure 5.2   Mechanism of epoxidation using titanium(IV) chiral complex.

We will describe representative procedures for the epoxidation of a disubstituted aromatic allylic alcohol (**A**), a trisubstituted aromatic allylic alcohol (**B**) and a disubstituted aliphatic allylic alcohol (**C**).

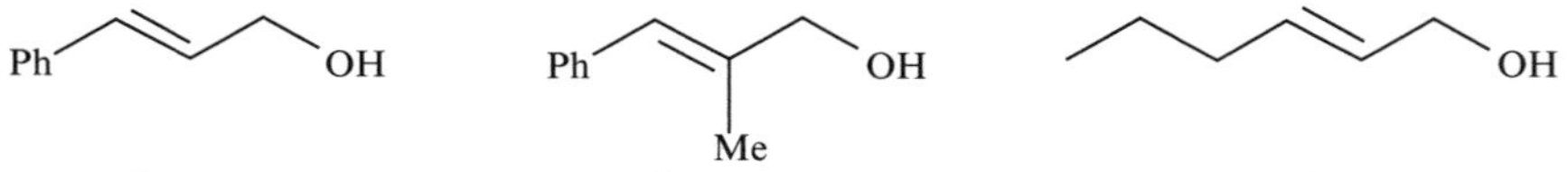

A   **B**   **C**

5.1   NON-ASYMMETRIC EPOXIDATION

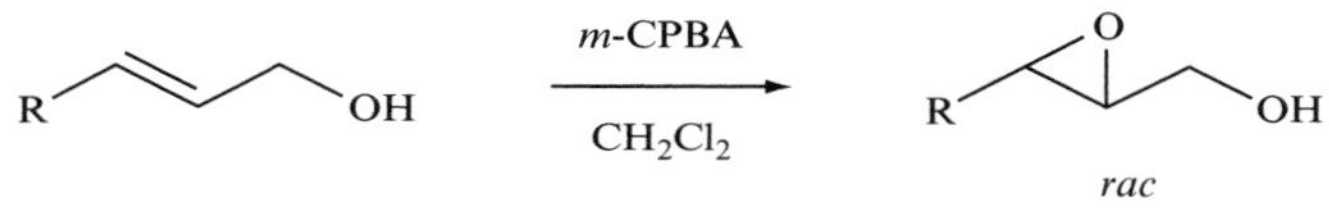

Materials and equipment

- Allylic alcohol, 1 mmol
- Anhydrous dichloromethane, 10 mL
- *m*-Chloroperbenzoic acid (MCPBA, *m*-CPBA), 1 mmol
- Saturated aqueous solution of sodium hydrogencarbonate, 40 mL
- Dichloromethane
- Magnesium sulfate
- Silica gel 60 (0.063–0.04 mm)

- 50 mL Round-bottomed flask with a magnetic stirrer bar
- Magnetic stirrer
- Ice-bath
- Separating funnel, 250 mL
- Rotary evaporator

Procedure

1. In a 50 mL round-bottomed flask was dissolved the allylic alcohol (1 mmol) in dry dichloromethane (10 mL). The mixture was cooled with an ice-bath, stirred, and *m*-chloroperbenzoic acid (1 mmol) was added.
2. The ice-bath was removed, the reaction mixture was stirred at room temperature and monitored by TLC. After completion of the reaction dichloromethane (10 mL) was added.
3. The reaction mixture was transferred into a separating funnel. The aqueous layer was extracted with dichloromethane (10 mL). The combined organic layers were washed with a aqueous solution of sodium hydrogencarbonate (2 × 20 mL), then with water (30 mL), dried over magnesium sulfate, filtered and the solvent removed under reduced pressure.
4. The residue was purified by flash chromatography over silica gel.
   See below for the method of purification for each product.

5.2 ASYMMETRIC EPOXIDATION USING A CHIRAL TITANIUM COMPLEX

The following three procedures need to be carried out under strictly anhydrous conditions.

- Before each reaction, the molecular sieves (4 Å in powder form or 3 Å as pellets) were activated by heating for 2 hours at 400 °C, then cooled under vacuum in a desiccator.
- Dry dichloromethane was stored on preactivated molecular sieves 3 Å in pellets (4 Å sieves should not to be used).
- The tartrate esters can be used as obtained from Aldrich Chemical Co. or Fluka Chemical Corp. If the yield and/or the enantiomeric excess is/are

lower than expected, the reaction should be repeated with tartrate distilled under high vacuum and stored under vacuum or in an inert atmosphere.

- An anhydrous solution of 5.5 M of *tert*-butyl hydroperoxide in *iso*octane stored over molecular sieves is available from Fluka.
- Liquid allylic alcohols ((E)-2-methyl-3-phenyl-2-propenol and (E)-2-hexen-1-ol) were stored over preactivated 3 Å molecular sieves.
- Titanium *iso*propoxide needs to be manipulated carefully with gloves and eye protection. If the yield and/or the enantiomeric excess is/are lower than expected, the catalyst should be distilled under vacuum (b.p 78–79.5 °C, 1.1 mmHg).

5.2.1 EPOXIDATION OF CINNAMYL ALCOHOL[3,4]

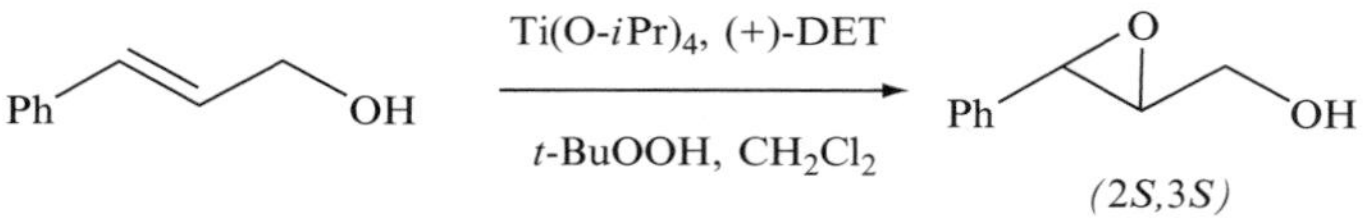

$$\text{Ph}\diagup\!\!\!\diagup\!\!\!\diagdown\text{OH} \xrightarrow[\text{t-BuOOH, CH}_2\text{Cl}_2]{\text{Ti(O-iPr)}_4,\ (+)\text{-DET}} \text{Ph}\overset{O}{\diagdown\!\!\!\diagup\!\!\!\diagdown}\text{OH}$$
(2S,3S)

Materials and equipment

- L-(+)-Di*iso*propyl tartrate ((+)-DIPT), 400 mg, 2 mmol, 0.1 eq
- Dichloromethane stored over preactivated 3 Å molecular sieves, 43 mL
- Activated powdered 4 Å molecular sieves, 500 mg
- Titanium isopropoxide, 297 μL, 1 mmol, 0.05 eq
- Anhydrous solution of 5.5 M of *tert*-butyl hydroperoxide in isooctane stored over molecular sieves, 5.5 mL, 30 mmol, 1.5 eq
- Cinnamyl alcohol, 2.68 g, 20 mmol
- Aqueous solution of sodium hydroxide 30 % saturated with sodium chloride, 6 mL
- Celite®
- Brine
- Magnesium sulfate
- Silica gel 60 (0.063–0.04 mm)
- (R)-(+)-α-Methoxy-α-(trifluoromethyl)phenylacetyl chloride (MTPA chloride) or the (S)-enantiomer, 5 mg, 0.02 mmol
- 4-Dimethylaminopyridine (DMAP), 5 mg, 0.04 mmol
- Dichloromethane, diethyl ether, petroleum ether, ethyl acetate, *n*-hexane
- *p*-Anisaldehyde

- 50 mL Two-necked flask with a magnetic stirrer bar
- Magnetic stirrer
- Cooling bath (acetone/Dri-ice) equipped with contact thermometer, −5 °C
- Büchner funnel with glass frite (30 mL, porosity n°3)

- Syringes
- Separating funnel, 250 mL
- Rotary evaporator

Procedure

1. A 50 mL two-necked flask equipped with a stirrer bar was placed in an oven at 120 °C overnight, cooled under vacuum and flushed with nitrogen.
2. The flask was filled with activated powdered 4 Å molecular sieves (500 mg), dry dichloromethane (40 mL) and L-(+)-diisopropyl tartrate (400 mg).
3. The mixture was cooled to −5 °C with the cooling bath, stirred and titanium isopropoxide (297 µL) was added. After cooling the bath to −20 °C, a solution of *tert*-butyl hydroperoxide (5.5 M in *iso*octane, 5.5 mL) was added and the mixture was stirred at −20 °C for 1 hour.
4. The solution of cinnamyl alcohol (2.68 g in 3 mL of dry dichloromethane) was added dropwise over 1 hour via a syringe.
5. The reaction was monitored by TLC (eluent: petroleum ether–diethyl ether, 1:1). The visualization of the cinnamyl alcohol (UV active) with *p*-anisaldehyde dip gave a blue stain, R_f 0.35, and a brown stain for the epoxycinnamyl alcohol, R_f 0.25.
6. After being stirred for 2 hours at −15 °C, the reaction was quenched with water (6 mL) and the mixture was stirred for 30 minutes at this temperature. The solution was warmed to room temperature. Hydrolysis of the tartrate was then effected by adding an aqueous solution of sodium hydroxide (30 %) saturated with sodium chloride (6 mL) and stirring vigorously for 1 hour.
7. A Büchner funnel with glass frite was packed with 2 cm Celite®. The two-phase mixture was filtered over the pad of Celite®, transferred into a separating funnel and the organic layer was separated.
8. The aqueous phase was washed with dichloromethane (3 × 10 mL) and the combined organic phases were dried (magnesium sulfate) and evaporated under reduced pressure to afford crude product.
9. The crude material was purified by flash chromatography over silica gel (150 g) using ethyl acetate-*n*-hexane (1:9) as eluent to give (2S,3S)-2,3-epoxy-3-phenyl-1-propanol as a white solid (2.08 g, 70 %).

    The ee (92 %) was determined by HPLC analysis (Chiralpak® OD column, flow 1 mL/min, isopropanol–*n*-hexane; 1:9); (2R,3R)-enantiomer: R_t 12.3 min, (2S,3S)-enantiomer: R_t 13.4 min or by analysis of the corresponding Mosher esters.

Derivatization with MTPA chloride

Esterification of chiral alcohols with (R)-(+)-α-methoxy-α-(trifluoromethyl)-phenylacetyl chloride (MTPA chloride) or its (S)-enantiomer as homochiral auxiliaries affords the corresponding diastereoisomeric (R)- or (S)-Mosher esters, respectively.

In a NMR tube, to a solution of the epoxy alcohol (2.5 mg) in $CDCl_3$ (0.5 mL) was added 4-dimethylaminopyridine (5 mg) and (R)-(+)-α-methoxy-α-(trifluoromethyl)phenylacetyl chloride (5 mg). The mixture was allowed to stand overnight at room temperature. The reaction was monitored by TLC to ensure complete consumption of the starting material. ^{1}H and ^{19}F NMR spectra were carried out on the crude reaction mixture. In the ^{19}F NMR spectrum, each enantiomer gave a signal; an additional signal at -71.8 ppm was ascribed to residual MTPA. (^{19}F NMR (250 MHz, $CDCl_3$): $\delta - 70.7$ (s, (2R,3R)-enantiomer); -72.0 (s, (2S,3S)-enantiomer)).

^{1}H NMR (200 MHz, $CDCl_3$): δ 7.37–7.27 (m, 5H, Ph); 4.06 (ddd, J 10.5 Hz, J 4.9 Hz, J 3.0 Hz, CH–CH$_2$); 3.94 (d, J 3 Hz, 1H, CH–Ph); 3.81 (dd, J 12.5 Hz, J 4.9 Hz, 1H, CHaHb), 3.24 (m, 1H, CHaHb), 2.05 (br, 1H, OH).

IR ($CHCl_3$, cm^{-1}): 3603, 3450 (O–H), 3060 (C–H epoxide), 3010 (C–H aromatic), 2927, 2875 (C–H aliphatic), 1606, 1461 (C=C), 1390, 1308, 1289, 1203 (C–OH, C–O–C), 1103, 1079, 1023, 882, 861, 838.

Mass: calculated for $C_9H_{10}O_2$: m/z 150.06808; found $[M]^{+\circ}$ 150.06781.

5.2.2   EPOXIDATION OF (E)-2-METHYL-3-PHENYL-2-PROPENOL[4]

Materials and equipment

- L-(+)-Diisopropyl tartrate ((+)-DIPT), 350 mg, 1.5 mmol, 0.075 eq
- Dichloromethane stored over preactivated 3 Å molecular sieves, 50 mL
- Activated powdered 4 Å molecular sieves, 1.2 g
- Titanium isopropoxide, 297 μL, 1 mmol, 0.05 eq
- Anhydrous solution of 5.5 M of tert-butyl hydroperoxide in isooctane stored over molecular sieves, 7.2 mL, 40 mmol, 2 eq
- (E)-2-Methyl-3-phenyl-2-propenol, 3 g, 2.9 mL, 20 mmol
- Aqueous solution of sodium hydroxide (30%) saturated with sodium chloride, 6 mL
- Celite®
- Brine
- Sodium sulfate
- Silica gel 60 (0.063–0.04 mm)
- Dichloromethane, diethyl ether, petroleum ether, methanol
- p-Anisaldehyde

- 50 mL Two-necked flask with a magnetic stirrer bar
- Magnetic stirrer
- Acetone/Dry ice cooling bath equipped with contact thermometer, $-35\,°C$
- Syringes
- Büchner funnel with glass frite, 30 mL
- Glass wool
- Separating funnel, 250 mL
- Rotary evaporator

Procedure

1. A 50 mL two-necked flask equipped with a stirrer bar was placed in an oven at $120\,°C$ overnight, cooled under vacuum and flushed with nitrogen.
2. The flask was filled with dry dichloromethane (50 mL) and L-(+)-diisopropyl tartrate (350 mg). The mixture was cooled to $-35\,°C$ using the cooling bath, then activated powdered 4 Å molecular sieves (1.2 g), titanium isopropoxide (297 µL) and a solution of tert-butyl hydroperoxide (5.5 M in isooctane, 7.2 mL) were added sequentially. The mixture was stirred at $-35\,°C$ for 1 hour.
3. The (E)-2-methyl-3-phenyl-2-propenol (2.9 mL) was added dropwise via a syringe over 30 minutes.
4. The reaction was monitored by TLC (eluent: petroleum ether–diethyl ether, 6:4). 2-Methyl-3-phenyl-2-propenol (UV active) visualized with a p-anisaldehyde dip stained blue, R_f 0.50 and the epoxide stained brown, R_f 0.33.
5. The mixture was stirred for 2.5 hours at $-35\,°C$, then the bath was warmed to $0\,°C$ and the reaction quenched by the addition of water (6 mL). The resulting mixture was stirred for 30 minutes.
6. The solution was warmed to room temperature. Hydrolysis of the tartrate was then effected by adding an aqueous solution of sodium hydroxide (30%) saturated with sodium chloride (6 mL) and stirring vigorously for 1 hour.
7. The mixture was transferred into a separating funnel. The aqueous phase was extracted with dichloromethane (2×30 mL). Then the combined organic layer phase were dried over sodium sulfate, filtered and concentrated under reduced pressure to give a yellow oil.

    If the phase separation does not occur, the reaction mixture is transferred into a separating funnel. A small amount of methanol is added to the mixture (5 mL) followed by a very brief shaking. Immediate phase separation often occurs, allowing for the simple removal of the lower organic phase. If the emulsion is still a problem, then the mixture is filtered once or twice through a small plug of glass wool washed with dichloromethane.
8. The crude material was purified by flash chromatography over silica gel (100 g) using petroleum ether–diethyl ether (8:2) as eluent to give (2S,3S)-2-methyl-3-phenyloxiranemethanol as a white solid (3 g, 18.6 mmol, 93%).

    The ee (94.5%) was determined by HPLC (Chiralpak® OD column, flow 1 mL/min, ethanol–n-hexane; 1:99); (2S,3S)-enantiomer: R_t 16.0 min,

($2R,3R$)-enantiomer: R_t 13.5 min. The analysis of the ester derived from (+)-MTPA chloride did not give any resolution by ^{19}F NMR.

^{1}H NMR (200 MHz, CDCl$_3$): δ 7.23–7.34 (m, 5H, Ph); 4.19 (s, 1H, CH); 3.68–3.88 (m, 2H, CH_2); 1.84–1.77 (m, 1H, OH); 1.06 (s, 3H, CH_3).

IR (CHCl$_3$, cm^{-1}): 3589, 3450, (O–H), 3060 (C–H epoxide), 3010 (C–H aromatic), 2938, 2878 (C–H aliphatic), 1713, 1603, 1493, 1452 (C=C), 1385, 1204 (C–OH, C–O–C), 1095, 1058, 1029, 980, 924, 898, 850.

5.2.3   EPOXIDATION OF (E)-2-HEXEN-1-OL[4]

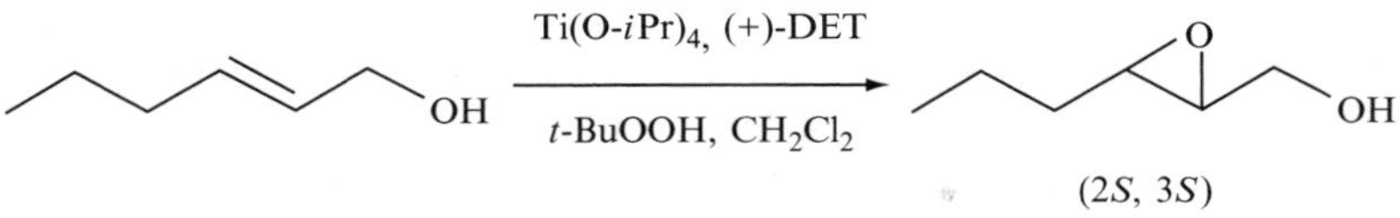

$$ \xrightarrow[\text{t-BuOOH, CH}_2\text{Cl}_2]{\text{Ti(O-}i\text{Pr)}_4, \ (+)\text{-DET}} $$

(2S, 3S)

Materials and equipment

- L-(+)-Diethyl tartrate ((+)-DET), 250 mg, 1.2 mmol, 0.06 eq
- Dichloromethane stored over preactivated 3 Å molecular sieves, 40 mL
- Activated powdered 4 Å molecular sieves, 600 mg
- Titanium isopropoxide, 297 μL, 1 mmol, 0.05 eq
- Anhydrous solution of 5.5 M of $tert$-butyl hydroperoxide in isooctane stored over molecular sieves, 7.2 mL, 40 mmol, 2 eq
- (E)-2-Hexen-1-ol, 2 g, 20 mmol
- Solution of ferrous sulfate heptahydrate, 6.6 g, 24 mmol, and tartaric acid, 2 g, 12 mmol, in deionized water, 20 mL
- Sodium hydroxide solution in saturated brine, 30%, 50 mL
- Sodium sulfate
- Silica gel 60 (0.063–0.04 mm)
- (R)-(+)-α-Methoxy-α-(trifluoromethyl)phenylacetyl chloride (MTPA chloride) or the (S)-enantiomer, 5 mg, 0.02 mmol
- 4-Dimethylaminopyridine (DMAP), 5 mg, 0.04 mmol
- n-Hexane, ethyl acetate, diethyl ether, triethylamine
- p-Anisaldehyde

- 50 mL Two-necked flask with a magnetic stirrer bar
- Magnetic stirrer
- Acetone/Dri-ice cooling bath equipped with contact thermometer, −20 °C
- Syringes
- Beaker, 100 mL
- Separating funnel, 250 mL
- Rotary evaporator

Procedure

1. A 50 mL two-necked flask equipped with a stirrer bar was placed in an oven at 120 °C overnight, cooled under vacuum and flushed with nitrogen.
2. To the flask was added dry dichloromethane (30 mL), activated powdered 4 Å molecular sieves (600 mg) and L-(+)-diethyl tartrate (250 mg).
3. After the mixture was cooled to −20 °C, titanium isopropoxide (297 μL) was added. The reaction mixture was stirred at −20 °C as a solution of *tert*-butyl hydroperoxide (5.5 M in *iso*octane, 7.2 mL) was added via a syringe at a moderate rate (over 5 minutes). The mixture was stirred at −20 °C for 30 minutes.
4. The solution of (E)-2-hexen-1-ol (2 g) in dry dichloromethane (10 mL) was added dropwise via a syringe over a period of 20 minutes, while the temperature was maintained between −20 °C and −15 °C.
5. The reaction mixture was stirred for an additional 2.5 hours at −20 °C. The reaction was monitored by TLC (eluent: *n*-hexane–ethyl acetate, 7:3). The products were visualized with a *p*-anisaldehyde dip; 2-hexenol stained purple, R_f 0.49 and the epoxide stained dark blue, R_f 0.22.
6. After completion of the reaction a 100 mL beaker containing the solution of ferrous sulfate–tartaric acid (20 mL) was pre-cooled at 0 °C by means of an ice-water bath. The epoxidation reaction mixture was allowed to warm to 0 °C and then was poured slowly onto the pre-cooled, stirring ferrous sulfate solution. The two-phase mixture was stirred for 5–10 minutes; the aqueous layer became brown.
7. The mixture was transferred into a separating funnel. The phases were separated and then the aqueous phase was extracted with diethyl ether (2 × 30 mL). The combined organic layers were treated with the pre-cooled solution of 30 % sodium hydroxide in saturated brine (50 mL).
8. The two-phase mixture was stirred vigorously for 1 hour at 0 °C and then diluted with 50 mL of water. The mixture was transferred into a separating funnel and the phases were separated. The aqueous layer was extracted with diethyl ether (2 × 50 mL) and the combined organic layers dried over sodium sulfate, filtered and concentrated under reduced pressure yielding a colourless oil.

   This procedure works well for most hydrophobic epoxy alcohols. The key advantage is that it is possible to remove tartrate, titanium isopropoxide and *tert*-butyl hydroperoxide, as those different compounds are not easily separated through distillation or recrystallization.
9. The crude material was purified by flash chromatography over silica gel (100 g), buffered with 1 % triethylamine, using *n*-hexane–diethyl ether (3:1) as eluent to give (2S,3S)-3-propyloxiranemethanol as a colourless oil (2 g, 15.3 mmol, 80 %).

   The ee (93 %) was determined by GC analysis (Lipodex® E, 25 m, 0.25 mm ID, temperatures: column 70 °C isotherm, injector 250 °C, detector 250 °C, mobile phase helium); (2S,3S)-enantiomer: R_t 53.6 min,

(2R, 3R)-enantiomer: R_t 52.6 min. The ee can be determined by analysis of the ester derived from (+)-MTPA chloride (^{19}F NMR (250 MHz, CDCl$_3$): δ − 70.8 (s, (2R,3R)-enantiomer); −72.0 (s, (2S,3S)-enantiomer)).

^{1}H NMR (200 MHz, CDCl$_3$): δ 3.91 (d, J 13.5 Hz, 1H); 3.60 (dd, J 13.4 Hz, J 4.1 Hz, 1H); 2.94 (m, 2H); 2.53 (m, 1H); 1.53 (m, 4H); 0.96 (t, J 7.1 Hz, 3H, CH_3).

IR (CHCl$_3$, cm^{-1}): 3589 (C–O), 3009, 2965, 2937, 2877 (C–H), 1458 (C–H, CH$_3$), 1382, 1203 (C–OH, C–O–C), 1095, 1030, 970, 924, 897, 848.

Table 5.1 Catalytic asymmetric epoxidation of allylic alcohols using a combination of titanium *iso*propoxide. enantiomerically pure tartrate ester ((+)-DET or (+)-DIPT) and *tert*-butyl hydroperoxide (yield and enantiomeric excess, according to the relevant publication)[4].

	Yield %	ee % (configuration)
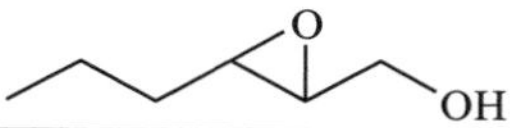	85*	94 (2S,3S)*
	89*	>98 (2S,3S)*
	74	86 (2S,3R)
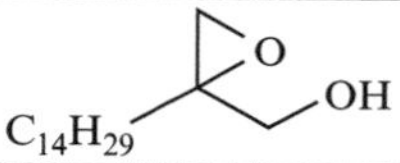	91	96 (S)
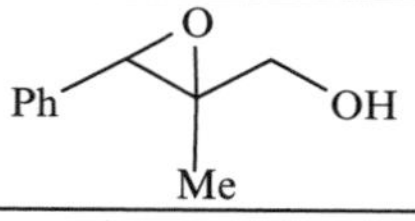	79*	>98 (2S,3S)*
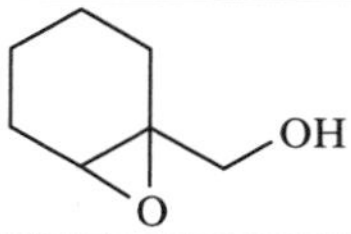	77	93
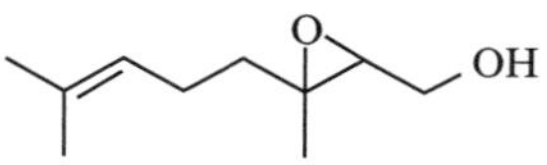	95	91 (2S,3R)
	70	91 (2S,3S)

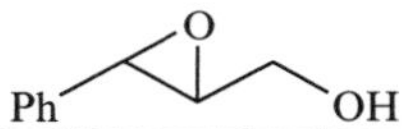

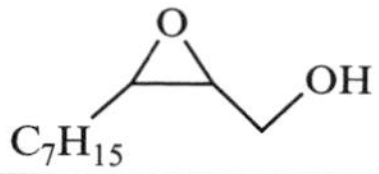

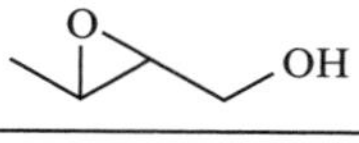

* Reaction described above

5.2.4 CONCLUSION

This method, specific for the epoxidation of allylic alcohols, gives good results if the reaction is carried out under strictly anhydrous conditions, otherwise the yield or the enantiomeric excess can decrease, sometimes dramatically. This can explain the small differences between the results obtained during the validation experiments and the published results. All the different reagents are commercially available; they can be used as received but in case of low yield and/or enantiomeric excess they should be distilled and dried under an inert atmosphere. Table 5.1 gives some other examples of substrates which can be epoxidized using the procedure described above.

5.3 ASYMMETRIC EPOXIDATION OF (*E*)-UNDEC-2-EN-1-OL USING POLY(OCTAMETHYLENE TARTRATE)

D.C. SHERRINGTON, J.K. KARJALAINEN and O.E.O. HORMI

Department of Pure and Applied Chemistry, Thomas Graham Building, University of Strathclyde, 295 Cathedral Street, Glasgow G1 1XL, Scotland, Tel: +44(0)1415482799, m.p.a.smith@strath.ac.uk

5.3.1 SYNTHESIS OF BRANCHED POLY(OCTAMETHYLENE-L-(+)-TARTRATE)[5]

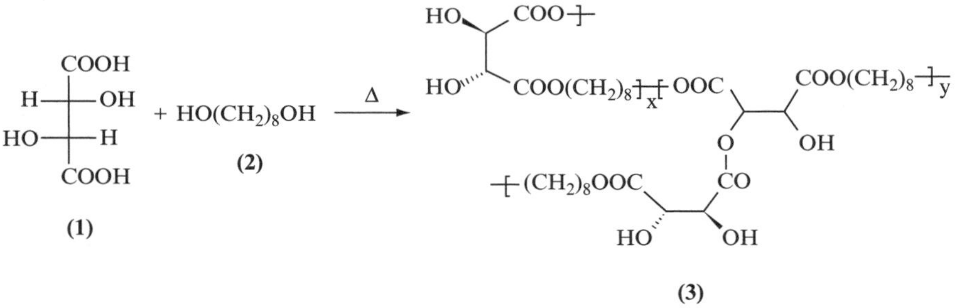

Materials and equipment

- L-(+)-Tartaric acid, 10.0 g, 0.067 mol
- 1,8-Octanediol, 11.7 g, 0.080 mol
- *p*-Toluene sulfonic acid, 0.6 g
- Ethyl acetate
- *n*-Hexane

- 100 mL Three-necked round-bottomed flask with a magnetic stirrer bar; N_2 cylinder and bubbler; oil bath; hot-plate stirrer; vacuum distillation equipment

Procedure

1. The tartaric acid, 1,8-octanediol and p-toluene sulfonic acid were placed in the flask and the latter flushed with N_2. A positive pressure of N_2 was then maintained throughout. The mixture was stirred as the temperature was raised to 140 °C to achieve a homogeneous solution; the temperature was then allowed to fall to 125 °C and the reaction left to proceed for 3 days.
2. Water and excess diol were removed by distillation under high vacuum to yield a solid mass. The latter was swollen in refluxing ethyl acetate (sufficient to make the mass mobile) and the resulting mixture poured into n-hexane ($\sim$3 fold volume excess).
3. The solid was recovered by decanting off the solvents and the polymer dried under vacuum at 40 °C for 6 hours and at room temperature for 2 days to yield 16.6 g (95%) of poly(octamethylene tartrate) (3).

    $[\alpha]_D^{25} = +9$ (c 1.6, THF).

    ^{1}H NMR (400 MHz, DMSO-d$_6$, 70 °C): δ 5.75 (br s), 5.41 (d, J 3.2 Hz), 4.62 (d, J 2.9 Hz), 4.37 (s, 2H), 4.09 (t, J 6.5 Hz, 4H), 1.58 (m, 4H), 1.30 (m, 9H).

    Note 1 – small signals δ = 5.41 and 4.62 correspond to methine H atoms on tartrate branch points; the ratio of these intensities to the total intensity of all tartrate methine resonances allows estimation of the percentage branching.

    Note 2 – the percentage branching can vary with precise reaction conditions, up to $\sim$ 10% gives optimal results; products insoluble in DMSO are crosslinked and should be discarded.

    FTIR (KBr, cm^{-1}) 3450 (OH), 2932, 2857 (C–H aliphatic), 1743 (C=O ester).

    Poly(octamethylene tartrate) can be used directly in place of dialkyl tartrates in the Sharpless epoxidation of allylic alcohols.

5.3.2   ASYMMETRIC EPOXIDATION OF (E)-UNDEC-2-EN-1-OL

R = C_3H_7, (**4a,5a**); C_8H_{17}, (**4b,5b**); Ph (**4c,5c**) (see Table 5.2)

Materials and equipment

- Dry CH_2Cl_2 (over $CaCl_2$)
- Powdered activated 4 Å molecular sieves

- Poly(octamethylene-L(+)-tartrate)
- Ti(OiPr)$_4$
- Anhydrous *tert*-butylhydroperoxide (TBHP) (in *iso*-octane).
  See refs 5.6 for preparation and standardization
- (E)-Undec-2-en-1-ol (**4**, R = C$_8$H$_{17}$)
- Diethyl ether, toluene, petroleum ether
- NaOH, NaCl, MgSO$_4$, Celite

- Three-necked round-bottomed flask
- Magnetic stirrer
- N$_2$ supply
- Gas bubbler
- Syringe
- Bücher funnel, flask
- Filter paper
- −20 °C Cold bath
- Rotary evaporator

Procedure

1. An oven-dried three-necked round-bottomed flask (100 mL) equipped with a magnetic stir bar, nitrogen inlet, septum and bubbler was charged with CH$_2$Cl$_2$ (35 mL dried over CaCl$_2$), powdered activated 4 Å molecular sieves (0.3 g) and poly(octamethylene-L-(+)-tartrate) (1.56 g, 0.0059 mol tartrate, 6% branching). The flask was cooled to −20°C and Ti(OiPr)$_4$ (0.85 g, 0.0030 mol) added via a syringe.

2. The mixture was stirred for 1 hour at −20 °C and then anhydrous TBHP (7.5 mL, 3.2 M in *iso*-octane) also added slowly via a syringe. The mixture was again stirred at −20 °C for 1 hour. (E)-Undec-2-en-1-ol (1 g, 0.0059 mol) in CH$_2$Cl$_2$ (5 mL) was added dropwise by syringe such that the temperature was maintained between −15 and −20 °C. The reaction mixture was stirred at −20 °C for 6 hours and then placed in a freezer overnight.

3. The polymer–ligand–Ti complex was filtered off and washed thoroughly with CH$_2$Cl$_2$. The recovered solution was then quenched with aqueous NaOH (30%, 10 mL, saturated with NaCl) and diethyl ether added (50 mL) after which the cold bath was removed and the stirred mixture allowed to warm up to 10 °C. Stirring was continued for 10 minutes at 10 °C whereupon sufficient magnesium sulfate and Celite were added to absorb all the aqueous phase. After a final 15 minutes stirring the mixture was allowed to settle and the solution filtered through a pad of Celite and washed with diethyl ether. The solvents were removed under vacuum and the excess TBHP removed by azeotropic distillation with a little added toluene. The crude product was purified by recrystallization from petroleum ether to yield a white solid (0.55 g, 50%).

- ee = 88 % determined by ^{1}H NMR analysis of the Mosher ester; $[\alpha]_D^{25} =$ −22(c 1.33, CHCl$_3$).
- ^{1}H NMR (200 MHz, CDCl$_3$): δ 3.88 (dd, $J2$, 13 Hz, 1H), 3.59 (ddd, $J5$, 7, 10 Hz, 1H), 2.87–2.96 (m, 2H), 2.68 (br s), 1.24–1.55 (m, 14H), 0.85 (t, J 7 Hz, 3H).
- FTIR (CHCl$_3$ cm^{-1}): 1237 (in phase epoxy); 933–817 (out of phase epoxy).

Utility and Scope

Use of poly(octamethylene tartrate) in place of dialkyl tartrates offers practical utility since the branched polymers yield *hetereogeneous* Ti complex catalysts which can be removed by filtration. Overall the work-up procedure is considerably simplified relative to the conventional Sharpless system. In addition, significant induction is shown in the epoxidation of (Z)-allylic alcohols[7] and even with homoallylic[8] species where the dialkyltartrates give very poor results Figure 5.3. Table 5.2 is illustrative of the scope using the polymer ligand.

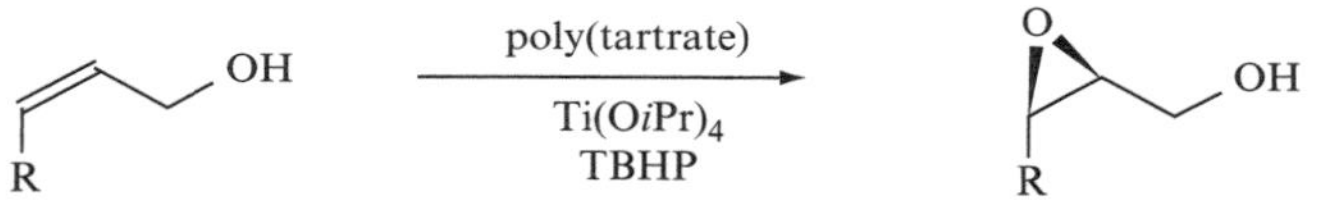

R=C$_2$H$_5$,(6a,7a); C$_3$H$_7$,(6b,7b);   PhCH$_2$OCH$_2$ (6c,7c)

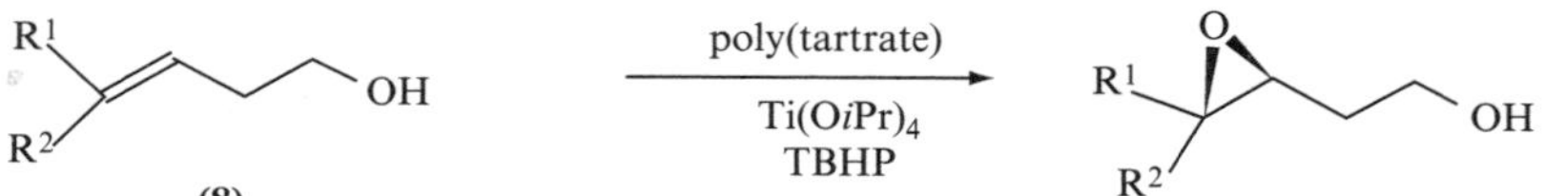

R^1= C$_2$H$_5$	R^2= H	(8a,9a)
R^1= H	R^2= C$_2$H$_5$	(8b,9b)
R^1= CH$_3$	R^2= CH$_3$	(8c,9c)
R^1= H	R^2= H	(8d,9d)

Figure 5.3 Oxidation of some (Z)-allylic alcohols and some homoallylic alcohols using poly(tartrate).

Table 5.2 Asymmetric epoxidation of *cis*- and *trans*-allylic and homoallylic alcohols using poly(octamethylene tartrate)/Ti(O*i*Pr)$_4$/TBHP.

Alkene	Epoxide	Poly(tartrate) % branching	Molar ratio alkene:Ti:tartrate	Temperature (°C)	Time	Isolated[a] Yield (%)	Ee(%)
4a	5a[b]	3	10:25:5	−20	6h	53	87[h]
4b	5b[b]	6	10:2:6	−15	12h	40	98[h]
4c	5cc	10	10:10:20	−20	7d	51	86[i]
6b	7b[c]	10	10:10:40	−20	6d	48	80[i]
6c	7c[c]	10	10:20:40	−20	6d	18	68[i]
8a	9a[d]	8	10:20:40	−20	5d	45	54[j]
8b	9b[e]	10	10:20:40	−20	21d	20	51[j]
8c	9c[f]	3	10:10:20	−20	1d	31	36[j]
8d	9d[g]	10	10:20:40	−20	14d	20	80[j]

a) GC yield typically much higher; scope for improvement in isolation
b) (2S-*trans*)isomer using L-(+)-polytartrate
c) (2S-*cis*)isomer using L-(+)-polytartrate
d) (3R, 4R)isomer using L-(+)-polytartrate
e) (3R, 4S)isomer using L-(+)-polytartrate
f) (3R)isomer using L-(+)-polytartrate
g) (3R)isomer using L-(+)-polytartrate
h) marginally lower than with dialkyltartrate
i) marginally better than with dialkyltartrate
j) substantially better than with dialkyltartrate

REFERENCES

1. Katsuki, T., Sharpless, K.B. *J. Am. Chem. Soc.*, 1980, **102**, 5974.
2. Berrisford, D.J., Bolm, C., Sharpless, K.B. *Angew. Chem. Int. Ed. English*, 1995, **34**, 1059.
3. Yang, Z.-C., Zhou, W.-S. *Tetrahedron*, 1995, **51**, 1429.
4. Gao, Y., Hanson, R.M., Klunder, J.M., Ko, S.Y., Masamune, H., Sharpless, K.B. *J. Am. Chem. Soc.*, 1987, **109**, 5765.
5. Karjalainen, J.K., Hormi, O.E.O., and Sherrington, D.C. *Tetrahedron Asymmetry*, 1998, **9**, 1563.
6. Hill, J.G., Rossiter, B.E., and Sharpless, K.B. *J. Org. Chem.*, 1983, **48**, 3607.
7. Karjalainen, J.K., Hormi O.E.O., and Sherrington, D.C. *Tetrahedron Asymmetry*, 1998, **9**, 2019.
8. Karjalainen, J.K., Hormi, O.E.O., and Sherrington, D.C. *Tetrahedron Asymmetry*, 1998, **9**, 3895.

6 Epoxidation of Unfunctionalized Alkenes and α, β-Unsaturated Esters

CONTENTS

The methods developed by E. Jacobsen[1], using the salen – manganese complexes, and Y. Shi[2], using chiral ketones, permit the epoxidation of a large range of disubstituted Z- or E-alkenes and trisubstituted alkenes. The methodology of Zhang, using porphyrins, is complementary.

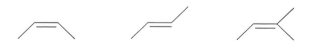

Disubstituted Z-alkene   Disubstituted E-alkene   Trisubstituted alkene

6.1 ASYMMETRIC EPOXIDATION OF DISUBSTITUTED Z-ALKENES USING A CHIRAL SALEN–MANGANESE COMPLEX[1]

Epoxidation of a variety of alkenes may be effected in a biphasic reaction system consisting of aqueous sodium perchlorate at pH $\geq$ 9.5 and an organic phase containing catalytic levels of a soluble manganese(III) complex (Figure 6.1). The ideal pH range appears to be 10.5–11.5 for most applications, with non-water-miscible solvents such as dichloromethane, *tert*-butyl methyl ether or ethyl acetate as the organic solvent. At pH $\leq$ 11.5 no phase transfer catalysts are necessary for epoxidation to occur, due to the presence of a significant equilibrium concentration of HOCl. At low pH, equilibrium levels of Cl_2 can produce chlorinated by-product. Reactions with alkenes are carried out in an air atmosphere, without the need to exclude moisture or trace impurities.

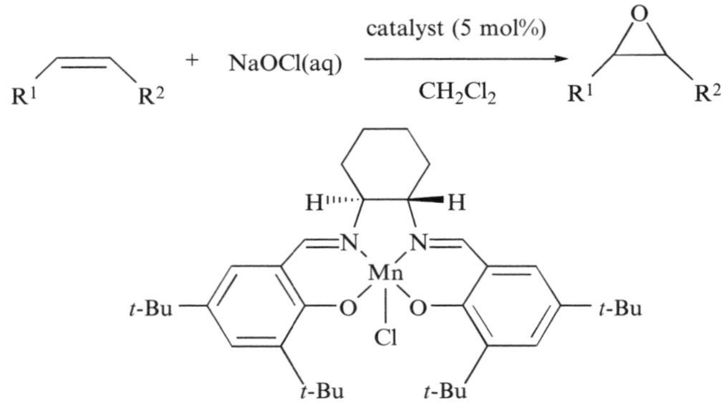

catalyst: (S,S)-(−)-N,N'-Bis(3,5-di-*tert*-butylsalicylidene)
-1,2-cyclohexanediaminomanganese(III) chloride

Figure 6.1   Epoxidation of Z-alkenes using a manganese(III) complex.

Mechanistically, the epoxidation appears to proceed via oxygen-atom transfer from the high-valent oxometallo intermediate (**A**) to organic substrates.

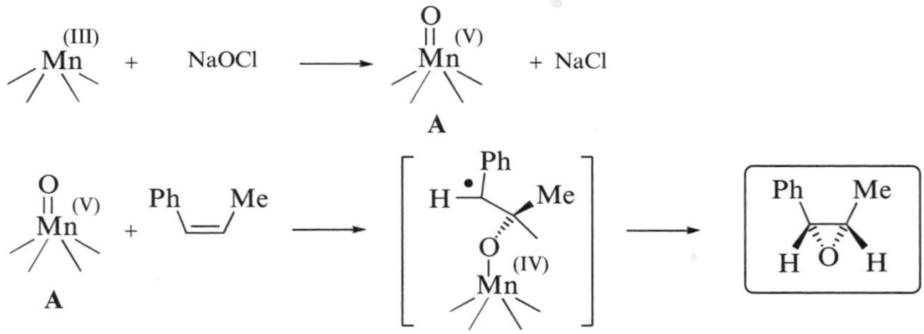

Figure 6.2   Mechanism of epoxidation using a manganese(III) complex.

There seems to be a direct attack of alkene at the oxometal, with C–O bond formation (Figure 6.2).

6.1.1 EPOXIDATION OF (Z)-METHYL STYRENE[3]

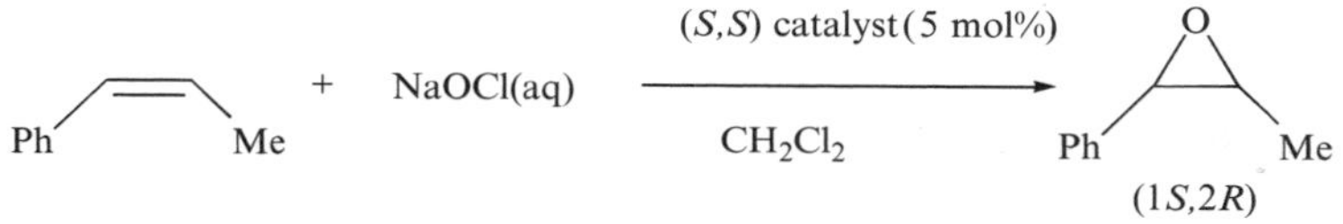

Materials and equipment

- Jacobsen's catalyst ((S,S)-(−)-N, N'-bis (3,5-di-*tert*-butylsalicylidene)-1,2-cyclohexanediaminomanganese(III) chloride, 98%), 26 mg, 0.04 mmol, 0.04 eq
- (Z)-Methyl styrene, 1 mmol
  (Z)-Methyl styrene was easily obtained by hydrogenation of 1-phenyl-1-propyne using Lindlar's catalyst (5% palladium on calcium carbonate, poisoned with lead) in *n*-hexane under an atmospheric pressure of hydrogen. The mixture, containing 90% of (Z)-methyl styrene and 10% of the over-reduced alkane, was used without further purification.
- Sodium hydrogenphosphate aqueous solution, 0.05 M, 5 mL
- Bleach (sodium hypochlorite, available chlorine 14%), 2.5 mL
- Sodium hydroxide solution, 1 M
- Hydrogen chloride solution, 1 M
- Dichloromethane, *n*-hexane, petroleum ether, diethyl ether, ethyl acetate, triethylamine
- Brine
- Sodium sulfate
- Silica gel 60 (0.063–0.04 mm)
- *p*-Anisaldehyde

- 100 mL Round-bottomed flask with a magnetic stirrer bar
- Magnetic stirrer
- Beaker, 100 mL
- Separating funnel, 100 mL
- Rotary evaporator
- pH-meter

Procedure

1. A solution was prepared by mixing an aqueous solution of sodium hydro-genphosphate (0.05 M, 5 mL) and a solution prepared with concentrated

bleach (sodium hypochlorite, 2.5 mL) in water (10 mL). The pH of the resulting solution was adjusted to pH 11.3 by addition of few drops of hydrogen chloride 1 M or sodium hydroxide 1 M. This solution was cooled using an ice-bath.

2. A 100 mL flask was filled with Jacobsen's catalyst (26 mg), (Z)-methyl styrene (1 mmol) and dichloromethane (1 mL). The solution was cooled with an ice-bath. To this solution was added the cold solution of bleach previously prepared (3.5 mL).

3. After 5 minutes the cooling bath was removed and the two-phase reaction mixture was stirred at room temperature. The reaction was monitored by TLC (eluent: petroleum ether–diethyl ether, 8:2). (Z)-Methyl styrene was UV active, R_f 0.85. The epoxide visualized with p-anisaldehyde dip stained yellow, R_f 0.63.

4. After 2 hours the reaction was quenched. The reaction mixture was transferred into a 100 mL beaker, n-hexane (10 mL) was added and stirred for 10 minutes.

5. The mixture was transferred into a separating funnel. The lower aqueous phase was separated and extracted with n-hexane (4×20 mL). The brown combined organic layers were washed with water (3×30 mL) and twice with brine (2×30 mL), dried over sodium sulfate, filtered and concentrated under reduced pressure, giving a brown oil (180 mg).

In case of formation of an emulsion, this solution was mixed with the aqueous layer and extracted with n-hexane as explained in the earlier procedure.

6. The crude material was purified by flash chromatography on silica gel (20 g) buffered with 1 % triethylamine, using petroleum ether–ethyl acetate (9:1) as eluent to give the epoxide as a grey oil (104 mg, 0.78 mmol, 78 % yield).

The ee (90.5 %) was determined by GC analysis (Lipodex® E 25 m, 0.25 mm ID, temperatures: column 110 °C isotherm, injector 250 °C, detector 250 °C, mobile phase helium); (1S,2R)-enantiomer: R_t 7.8 min, (1R, 2S)-enantiomer: R_t 9.0 min.

^{1}H NMR (200 MHz, CDCl$_3$): δ 7.35–7.31 (m, 5H, Ph); 4.07 (d, J 4.4 Hz, 1H, CH); 3.35 (qd, J 5.5 Hz, J 4.4 Hz, 1H, CH–CH$_3$); 1.09 (d, J 5.5 Hz, 3H, CH$_3$).

IR (CHCl$_3$, cm^{-1}): 3091, 3069 (C–H epoxide), 3008 (C–H aromatic), 2972, 2934, 2877 (C–H aliphatic), 1730, 1604, 1495, 1451 (C=C), 1416, 1375, 1359, 1248 (C–OH, C–O–C), 1028, 1009, 955, 915, 851.

Mass: calculated for C$_9$H$_{14}$NO: m/z 152.10754; found [M + NH$_4$]$^+$ 152.10743.

6.1.2 EPOXIDATION OF (Z)-ETHYL CINNAMATE[4]

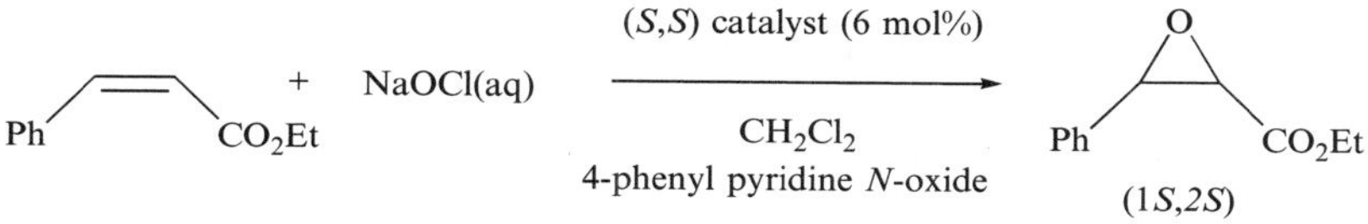

$$Ph \diagup \diagdown CO_2Et \quad + \quad NaOCl(aq) \quad \xrightarrow[\substack{CH_2Cl_2 \\ \text{4-phenyl pyridine } N\text{-oxide}}]{(S,S) \text{ catalyst (6 mol\%)}} \quad Ph \diagup\!\!\!\triangle\!\!\!\diagdown CO_2Et$$

(1S,2S)

Materials and equipment

- Jacobsen's catalyst (S,S)-$(-)$-N,N'-bis(3,5-di-*tert*-butylsalicylidene)-1,2-cyclohexane diaminomanganese(III) chloride, 98%), 108 mg, 0.166 mmol, 6.5 mol%
- (Z)-Ethyl cinnamate, 500 mg, 2.55 mmol

  The (Z)-ethyl cinnamate was obtained by hydrogenation of ethyl phenyl propiolate using Lindlar's catalyst (5% palladium on calcium carbonate, poisoned with lead) in *n*-hexane under an atmospheric pressure of hydrogen. The mixture, containing 75% of the (Z)-ethyl cinnamate, 22% of the over-reduced alkane and 3% of the (E)-ethyl cinnamate, was used without further purification. 666 mg of this mixture contains 500 mg of (Z)-ethyl cinnamate (2.55 mmol).
- 4-Phenylpyridine *N*-oxide, 116 mg, 0.680 mmol, 0.25 eq
- Sodium hydrogenphosphate aqueous solution 0.05 M, 10 mL
- Bleach (sodium hypochlorite, available chlorine 14%), 5 mL
- Sodium hydroxide solution, 1 M
- Hydrogen chloride solution, 1 M
- Dichloromethane, petroleum ether, diethyl ether, *tert*-butyl methyl ether, ethyl acetate
- Celite®
- Brine
- Sodium sulfate
- Silica gel 60 (0.063–0.04 mm)
- *p*-Anisaldehyde

- 100 mL Flask with a magnetic stirrer bar
- Magnetic stirrer
- Beaker, 100 mL
- Separating funnel, 250 mL
- Rotary evaporator
- pH-meter

Procedure

1. A solution was prepared by mixing an aqueous solution of sodium hydro-genphosphate ($0.05\,M$, $10\,\text{mL}$) and a solution prepared with concentrated bleach (sodium hypochlorite, $5\,\text{mL}$) in water ($20\,\text{mL}$). The pH of the resulting solution was adjusted to 11.25 by addition of few drops of hydrogen chloride $1\,M$ or sodium hydroxide $1\,M$.

2. A $100\,\text{mL}$ flask, was filled with (Z)-ethyl cinnamate, ($666\,\text{mg}$ of the mixture containing $75\,\%$ of (Z)-ethyl cinnamate), 4-phenylpyridine N-oxide ($116\,\text{mg}$) and dichloromethane ($6\,\text{mL}$). Jacobsen's catalyst ($108\,\text{mg}$) was then added.

3. The resulting organic solution and the buffered bleach ($16\,\text{mL}$) were cooled separately using an ice-bath and then combined at $4\,°\text{C}$ in the flask.

4. After 5 minutes the cooling bath was removed and the two-phase reaction mixture was stirred at room temperature. The reaction was monitored by TLC (eluent: petroleum ether–diethyl ether, 9:1). (Z)-Ethyl cinnamate was UV active, R_f 0.42. The epoxide visualized with p-anisaldehyde dip stained yellow, R_f 0.28.

5. After 12 hours of stirring the two-phase mixture at room temperature, the reaction was quenched. The reaction mixture was transferred into a $100\,\text{mL}$ beaker, *tert*-butyl methyl ether ($50\,\text{mL}$) was added and the solution was stirred for 10 minutes.

6. The mixture was transferred into a separating funnel. The aqueous phase was separated and extracted with *tert*-butyl methyl ether ($2 \times 20\,\text{mL}$). The brown combined organic layers were filtered through a pad of Celite®, and then washed with water ($2 \times 40\,\text{mL}$) and with brine ($2 \times 40\,\text{mL}$), dried over sodium sulfate, filtered and concentrated under reduced pressure, giving a brown oil ($700\,\text{mg}$).

7. The crude material was purified by flash chromatography on silica gel ($20\,\text{g}$) using petroleum ether–ethyl acetate (95:5) as eluent to give a yellow oil ($286\,\text{mg}$) which was a mixture of $77\,\%$ (Z)-ethyl 3-phenylglycidate ($1.14\,\text{mmol}$, $45\,\%$ yield) and $23\,\%$ of (E)-ethyl 3-phenylglycidate ($0.34\,\text{mmol}$).

    The ee ($91.5\,\%$) was determined by GC analysis (Lipodex® C $25\,\text{m}$, $0.25\,\text{mm}$ ID, temperatures: column $110\,°\text{C}$ isotherm, injector $250\,°\text{C}$, detector $250\,°\text{C}$, mobile phase helium). ($1R$, $2R$)-(Z)-enantiomer: R_t $41.4\,\text{min}$, ($1S,2S$)-(Z)-enantiomer: R_t $42.6\,\text{min}$.

    (Z)-Ethyl 3-phenylglycidate ^1H NMR ($200\,\text{MHz}$, $CDCl_3$): δ 7.42–7.30 (m, 5H, Ph); 4.27 (d, J $4.4\,\text{Hz}$, 1H, CH); 4.01 (q, J $7.15\,\text{Hz}$, 2H, CH_2); 3.83 (d, J $4.4\,\text{Hz}$, 1H, CH); 1.02 (t, J $7.15\,\text{Hz}$, 3H, CH_3).

    (E)-Ethyl 3-phenylglycidate ^1H NMR ($200\,\text{MHz}$, $CDCl_3$): δ 7.42–7.30 (m, 5H, Ph); 4.291 (q, J $7.15\,\text{Hz}$, 2H, CH_2); 4.09 (d, J $1.65\,\text{Hz}$, 1H, CH); 3.51 (d, J $1.65\,\text{Hz}$, 1H, CH); 1.33 (t, J $7.15\,\text{Hz}$, 3H, CH_3).

    IR ($CHCl_3$, cm^{-1}): 3070 (C–H epoxide), 3012 (C–H aromatic), 2988, 2943, 2911, 2876 (C–H aliphatic), 1748 (C=O), 1455 (C=C), 1416, 1375 (CH_2, CH_3), 1394, 1301 (C–O), 1190 (C–O–C)), 1108, 1052, 1027, 917.

    Mass: calculated for $C_{11}H_{12}O_3$: m/z 192.07864; found $[M]^{+\bullet}$ 192.07855.

6.1.3 CONCLUSION

Epoxidation using manganese – salen complexes is very easy to carry out; it occurs under aqueous conditions and commercial house bleach can be used as the oxidant. The results are similar to those reported in the literature; Table 6.1 gives other examples of alkenes which can be epoxidized using the same procedure. This method gives good results, especially for disubstituted Z-alkenes but trisubstituted alkenes can be epoxidized as well.

Table 6.1 Epoxidation of disubstituted Z-alkenes by (S,S)-salen–manganese complex (results according to the literature[1,5]).

	3. Yield%	ee %
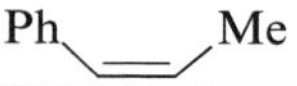 Ph⎯⎯Me	84*	91*
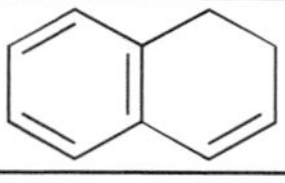 (naphthalene/dihydronaphthalene structure)	67	92
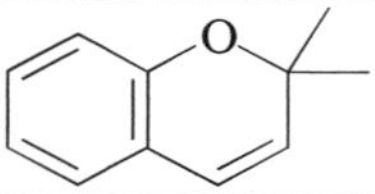 (2,2-dimethylchromene structure)	87	98
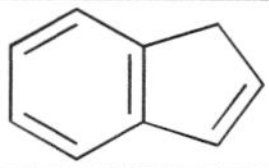 (indene structure)	80	88
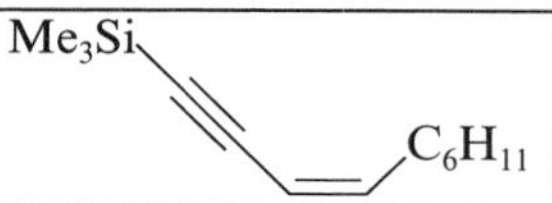 Me₃Si⎯⎯C₆H₁₁	65	98
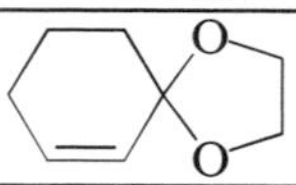 (cyclohexene dioxolane structure)	63	94
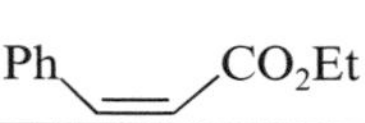 Ph⎯⎯CO₂Et	67*	97*
Trisubstituted alkene: 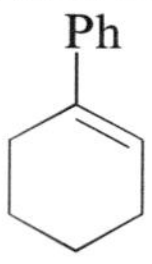 Ph (cyclohexene structure)	75	92

* Reaction described above

6.2 ASYMMETRIC EPOXIDATION OF DISUBSTITUTED *E*-ALKENES USING A D-FRUCTOSE BASED CATALYST[2]

Among many other methods for epoxidation of disubstituted *E*-alkenes, chiral dioxiranes generated *in situ* from potassium peroxomonosulfate and chiral ketones have appeared to be one of the most efficient. Recently, Wang *et al.*[2] reported a highly enantioselective epoxidation for disubstituted *E*-alkenes and trisubstituted alkenes using a D- or L-fructose derived ketone as catalyst and oxone as oxidant (Figure 6.3).

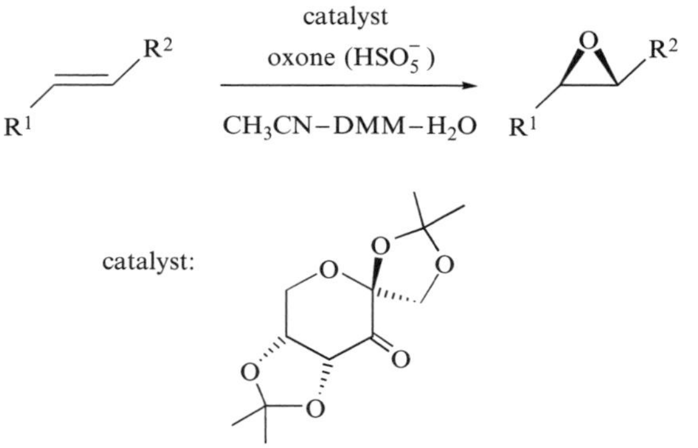

Figure 6.3 Epoxidation of *E*-alkenes by a ketone derived from D-fructose.

The ketone catalyst is readily prepared from D-fructose by ketalization and oxidation. The other enantiomer of this ketone, prepared from L-sorbose,

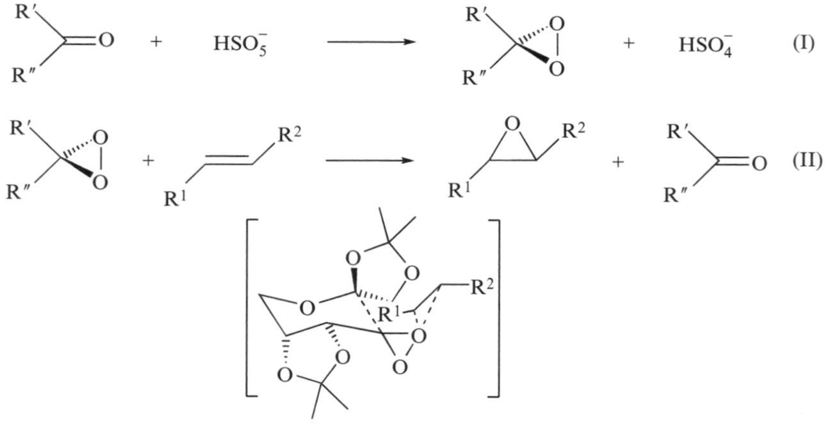

Figure 6.4 Mechanism of epoxidation by a chiral ketone catalyst.

shows the same enantioselectivity for the epoxidation[6]. The dioxirane is generated *in situ* from potassium monoperoxosulfate and the ketone catalyst (I, Figure 6.4). During the oxygen-atom transfer reaction from the dioxirane to the *E*-alkene (II, Figure 6.4), a 'spiro' transition state was proposed to give the (*E*)-epoxide.

For this method, all glassware needs to be carefully washed to be free of any trace of metals which catalyse the decomposition of oxone.

6.2.1   EPOXIDATION OF (*E*)-STILBENE

This reaction can be carried out at different temperatures (room temperature, −10 °C or −20 °C) but the reaction at room temperature gave a good compromise between the yield and the enantiomeric excess.

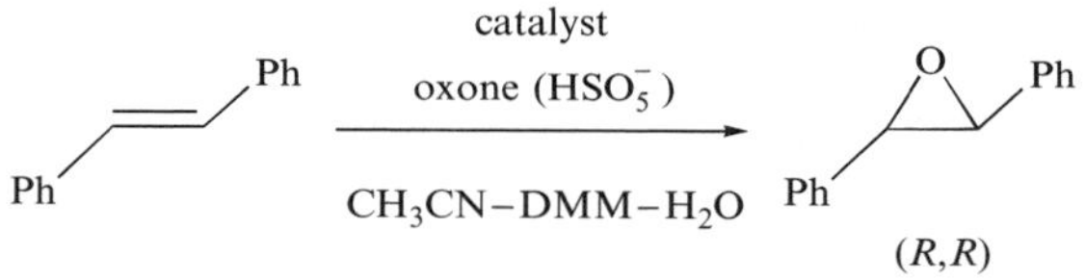

Materials and equipment

- (*E*)-Stilbene, 181 mg, 1 mmol
- Distilled water
- Aqueous solution of ethylenediamine tetraacetic acid disodium salt (Na$_2$ (EDTA)), 4×10^{-4} M, 20 mL
- Buffer: sodium tetraborate decahydrate 0.05 M (Na$_2$B$_4$O$_7$.10H$_2$O) in aqueous Na$_2$(EDTA)4×10^{-4} M, 10 mL
- Tetrabutylammonium hydrogensulfate, 15 mg, 0.04 mmol
- Ketone catalyst derived from fructose, 77.4 mg, 0.3 mmol, 0.3 eq*
- Solution of potassium peroxymonosulfate (oxone), 1 g, 1.6 mmol, 1.6 eq, in aqueous solution of 4×10^{-4} MNa$_2$(EDTA), 6.5 mL
- Solution of potassium carbonate, 930 mg, 6.74 mmol in water, 6.5 mL
- Acetonitrile (5 mL), dimethoxymethane (10 mL), pentane (5 mL), diethyl ether, *n*-hexane, triethylamine
- Brine
- Sodium sulfate
- Silica gel 60 (0.063–0.04 mm)
- *p*-Anisaldehyde

---

* *The ketone catalyst was kindly provided by Professor Y. Shi (Colorado State University, Fort Collins, Colorado)*

- 50 mL Three-necked flask with magnetic stirrer bar
- Magnetic stirrer
- Glass graduated cylinders
- Two addition funnels, 10 mL
- Separating funnel, 250 mL
- Rotary evaporator
- pH-meter

Procedure

1. In a 50 mL three-necked flask with a magnetic stirrer bar was dissolved (*E*)-stilbene (181 mg) in acetonitrile–dimethoxymethane (15 mL, 1/2, v/v). Buffer (10 mL), tetrabutylammonium hydrogensulfate (15 mg) and ketone catalyst (77.4 mg) were added with stirring.

    The mixture of organic solvent and borate buffer results in a solution of pH above 10, and with the addition of K_2CO_3 it rarely falls below that value.

2. The flask was equipped with two addition funnels; one of them was filled with the solution of oxone (1 g) in aqueous $Na_2(EDTA)$ (4×10^{-4} *M*, 6.5 mL) and the other one with a solution of potassium carbonate (930 mg) in distilled water (6.5 mL). The two solutions were added dropwise as slowly as possible over a period of 1 hour.

    To maximize the conversion and enantioselectivity a steady and uniform addition rate of oxone and K_2CO_3 must be achieved. On a small scale (1 mmol substrate), this is easily done with a syringe pump.

3. The reaction was monitored by TLC (eluent: *n*-hexane–diethyl ether, 9:1). (*E*)-Stilbene was UV active, R_f 0.82. The epoxide (UV active) stained blue with *p*-anisaldehyde, R_f 0.70.

4. After completion of the addition, the reaction was stirred for 1 hour and immediately quenched by addition of water (10 mL) and pentane (5 mL).

5. The reaction mixture was transferred into a separating funnel and was extracted with *n*-hexane (4×40 mL). The combined organic layers were washed with brine, dried over sodium sulfate, filtered and concentrated under reduced pressure to give a colourless oil (200 mg).

6. The crude material was purified by flash chromatography on silica gel (60 g), buffered with 1 % of triethylamine, using *n*-hexane–diethyl ether (95:5) to afford (*R,R*)-(*E*)-stilbene oxide as a colourless oil (123 mg, 0.62 mmol, 62 %).

    The ee (96 %) was determined by HPLC (Chiralpak® AD column, flow 1 mL/min, ethanol–*n*-hexane; 1:9); (*R,R*)-enantiomer: R_t 5.6 min, (*S,S*)-enantiomer: R_t 10.1 min.

    ^{1}H NMR (200 MHz, CDCl$_3$): δ 7.54–7.12 (m, 10H, Ph); 3.87 (s, 2H, CH).

    IR (CHCl$_3$, cm^{-1}): 3093, 3070 (C–H epoxide), 3012 (C–H aromatic), 2399, 1604, 1496, 1461, 1453 (C=C), 1230 (C–O–C), 1071, 1028, 925, 872, 838.

    Mass: calculated for $C_{14}H_{12}O$: *m/z* 196.08882; found $[M]^{+\bullet}$ 196.08861.

6.2.2 CONCLUSION

Epoxidation using a chiral fructose-derived ketone is easy to carry out, as it occurs in aqueous conditions. The reactions were performed without any modification of the published procedure. The glassware has to be free of trace metal, which can decompose the oxone; the use of a plastic spatula is recommended and the volumes must be measured using glass-graduated cylinders. Table 6.2 gives different examples of epoxides which can be obtained using the method prescribed.

Shi's method gives good results for disubstituted E-alkenes compared to the Jacobsen epoxidation previously described, which is more specific for disubstituted Z-alkenes. Concerning the epoxidation of trisubstituted alkenes, the epoxidation of 1-phenyl-1-cyclohexene could not be validated because of

Table 6.2 Epoxidation of disubstituted E-alkenes and trisubstituted alkenes by ketone derived from D-fructose[2].

	Yield %	ee % (configuration)
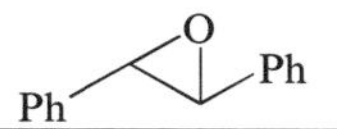	78*	>99 (R,R)*
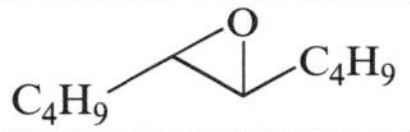	70	91 (R,R)
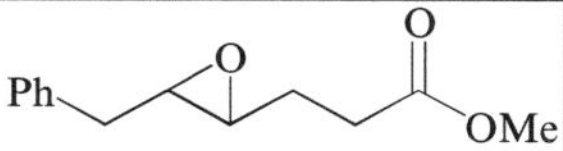	68	92 (R,R)
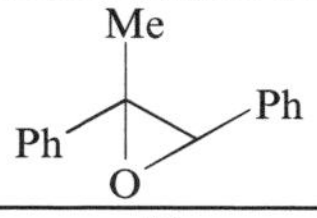	89	95.5 (R,R)
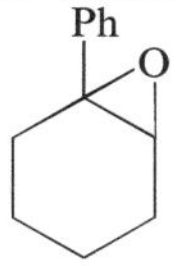	94	98 (R,R)
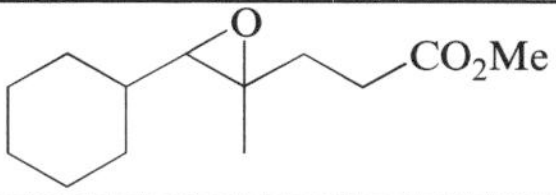	89	94 (R,R)
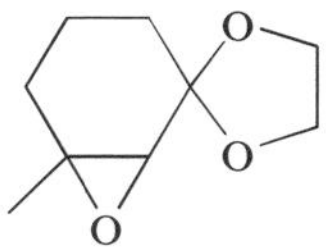	41	97.2 (R,R)

* Reaction described above

the difficulties in the determination of the enantiomeric excess; however, the yields were similar to the results given in the literature for both methods.

6.3   ENANTIOSELECTIVE EPOXIDATION OF (E)-β-METHYLSTYRENE BY D₂-SYMMETRIC CHIRAL *TRANS*-DIOXORUTHENIUM(VI) PORPHYRINS

Rui Zhang, Wing-Yiu Yu and Chi-Ming Che*

Department of Chemistry, The University of Hong Kong, Pokfulam Road, Hong Kong

6.3.1   PREPARATION OF THE *TRANS*-DIOXORUTHENIUM(VI) COMPLEXES WITH D₂ SYMMETRIC PORPHYRINS (H₂L$^{1-3}$)[7]

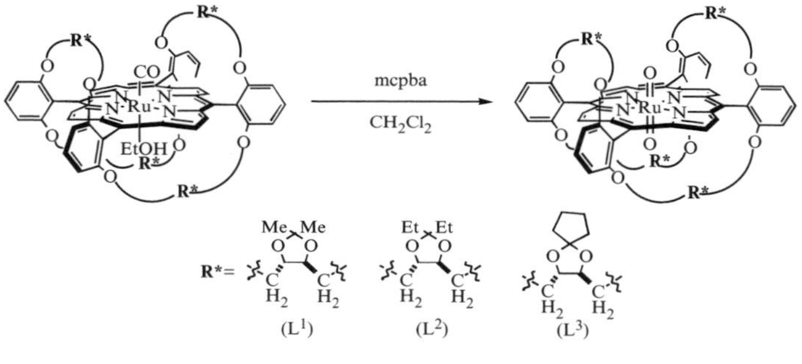

Materials and equipment

- Dichloromethane (freshly distilled over P_2O_5 under a nitrogen atmosphere), 5 mL
- *m*-Chloroperoxybenzoic acid (AR, Aldrich), 100 mg
- [Ru^{II}(L$^{1-3}$) (CO) (EtOH)], 0.07 mmol
- Acetone (AR)–CH₂Cl₂ (1:1), 20 mL
- Aluminum oxide (Brockman Grade II–III, basic), 20 g

- 10 mL Round-bottom flask with a magnetic stirrer bar
- Magnetic stirrer
- Vacuum pump

Procedure

1. A dichloromethane solution of [Ru^{II}(L$^{1-3}$) (CO) (EtOH)] (0.07 mmol) was added to a well-stirred dichloromethane solution of *m*-chloroperoxybenzoic acid (100 mg, 0.62 mmol) in a 10 mL round-bottomed flask. After 3 to 5 minutes stirring, the mixture was poured onto a short dry alumina column.

The completion of the oxidation can be checked by monitoring the disappearance and emergence of the Soret bands of the Ru(II) ($\lambda_{max} = 426$ nm) and the Ru(VI) ($\lambda_{max} = $ ca.440 nm) complexes using UV–vis spectroscopy. (Note: a prolonged reaction time (10 minutes) could result in lower product yields.)

2. The product complex was eluted by CH_2Cl_2/acetone (1:1 v/v). After solvent evaporation, a dark purple residue (~80 mg) was obtained. Yield: ~80 %.
3. The solid can be kept in a fridge (4 °C) for about one month without deterioration.

   $[Ru^{VI}(L^1)O_2]$ (**1a**): ^{1}H NMR (300 MHz, CDCl$_3$): δ 8.65 (d, J 4.7 Hz, 4H), 8.55 (d, J 4.7 Hz, 4H), 7.77 (t, J 6.5 Hz, 4H), 7.22–7.38 (m, meta-H overlapped with a solvent peak, 8H), 4.91 (d, J 10.4 Hz, 4H), 4.62 (d, J 9.0 Hz, 4H), 4.43 (t, J 9.0 Hz, 4H), 4.22 (d, J 10.1 Hz, 4H), 3.76 (d, J 9.0 Hz, 4H), 2.60 (t, J 8.5 Hz, 4H), 0.77 (s, 12H), −0.78 (s, 12H).

   IR (KBr): 818 ($\nu_{Ru=O}$), 1018 cm^{-1} (oxidation state marker band).

   UV–vis (CH_2Cl_2) λ_{max}/nm (log ε/dm^{3}mol^{-1}cm^{-1}): 442 (5.12), 536 (4.07).

   FAB–MS m/z: 1379 (M$^+$, 18 %), 1363 (M$^+$–O, 30 %), 1347 (M$^+$–2O, 100 %).

   $[Ru^{VI}(L^2)O_2]$ (**1b**): ^{1}H NMR (300 MHz, CDCl$_3$): δ 8.64 (d, J 4.7 Hz, 4H), 8.53 (d, J 4.7 Hz, 4H), 7.74 (m, 4H), 7.37–7.30 (m, meta-H overlapped with a solvent peak, 8H), 4.98 (d, J 10.4 Hz, 4H), 4.60 (d, J 8.8 Hz, 4H), 4.44 (t, J 8.9 Hz, 4H), 4.23 (d, J 10.4 Hz, 4H), 3.74 (d, J 8.8 Hz, 4H), 2.63 (t, J 8.7 Hz, 4H), 1.02 (m, 8H), 0.53 (t, J 7.2 Hz, 12H), −0.25 (m, 8H), −1.37 (t, J 7.3 Hz, 12H).

   IR (KBr): 821 ($\nu_{Ru=O}$), 1019 cm^{-1} (oxidation state marker band).

   UV–vis (CH_2Cl_2): λ_{max}/nm (log ε/dm^{3}mol^{-1}cm^{-1}) 443 (5.15), 534 (4.02).

   FAB–MS: m/z 1491 (M$^+$, 9 %), 1475 (M$^+$–O, 20 %), 1459 (M$^+$–2O, 100 %).

   $[Ru^{VI}(L^3)O_2]$ (**1c**): ^{1}H NMR (300 MHz, CDCl$_3$): δ 8.67 (d, J 4.6 Hz, 4H), 8.56 (d, J 4.6 Hz, 4H), 7.76 (m, 4H), 7.20–7.37 (m, meta-H overlapped with a solvent peak, 8H), 4.88 (d, J 9.3 Hz, 4H), 4.60 (d, J 9.4 Hz, 4H), 4.64 (t, J 9.0 Hz, 4H), 4.24 (d, J 10.1 Hz, 4H), 3.77 (d, J 8.6 Hz, 4H), 2.57 (t, J 8.6 Hz, 4H), 0.83 (m, 12H), 0.60 (m, 12H), −0.32 (m, 4H), −1.16 (m, 4H).

   IR (KBr): 819 ($\nu_{Ru=O}$), 1019 cm^{-1} (oxidation state marker band).

   UV–vis (CH_2Cl_2) λ_{max}/nm (log ε/dm^{3}mol^{-1}cm^{-1}): 441 (5.03), 535 (4.02).

   FAB–MS m/z: 1483 (M$^+$, 8 %), 1467 (M$^+$–O, 22 %), 1451 (M$^+$–2O, 100 %).

6.3.2 ENANTIOSELECTIVE EPOXIDATION OF (E)-β-METHYLSTYRENE

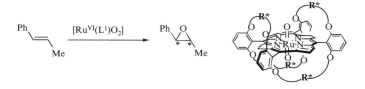

Materials and equipment

- Benzene (freshly distilled over sodium/benzophenone under a nitrogen atmosphere), 5 mL
- (E)-β-Methylstyrene (purified by simple distillation), 118 mg, 1 mmol
- [Ru^{VI}(L^1)O$_2$], 20–40 mg, 0.015–0.03 mmol
- Pyrazole (Hpz) (AR, Aldrich), 30 mg, 0.4 mmol
- Petroleum ether (AR) and CH$_2$Cl$_2$ (AR)
- Silica gel (70–230 mesh ASTM)

- 10 mL Round-bottomed flask with a magnetic stirrer bar
- Magnetic stirrer plate
- Glass column for chromatography
- Rotary evaporator
- HP-UV 8543 Ultraviolet–visible spectrophotometer
- HP5890 Series II Gas Chromatograph equipped with a chiraldex G-TA capillary column and a flame ionization detector

Procedure

1. In an ice-cooled (0 °C) 10 mL round-bottomed flask equipped with a magnetic stirrer bar were placed dry benzene (5 mL), (E)-β-methylstyrene (118 mg) and pyrazole (30 mg) under an argon atmosphere. [Ru^{VI}(L^1)O$_2$] (20–40 mg) was added to the mixture with stirring. The resulting solution was stirred for overnight at 0 °C under an inert atmosphere.

2. The reaction was monitored by a UV–vis spectrophotometer, and the completion of the reaction was confirmed by the disappearance of the Soret band at 442 nm.

3. The mixture was passed through a silica gel column and eluted initially with petroleum ether to remove the unreacted alkene. The product epoxide was collected by using CH$_2$Cl$_2$ as the eluent.

   The epoxide and aldehyde were identified and quantified by capillary GLC equipped with a Chiraldex G-TA column using 4-bromochlorobenzene as the internal standard (oven temp: 110 °C, carrier gas: He, flow rate: 60–65 mL min^{-1}, split ratio 100:1, detector: FID at 250 °C, (1S, 2S)-(E)-β-methylstyrene oxide: R$_t$ = 6.98 min; (1R, 2R)-(E)-β-methylstyrene oxide: R$_t$ = 7.81 min.) The enantiopurity of the 1R,2R-epoxide = 70% ee, Yield = 90% (>99% trans).

6.3.3 CONCLUSION

A D_2-symmetric chiral *trans*-dioxoruthenium(VI) porphyrin, [Ru^{VI}(L^1)O$_2$], bifacially encumbered by four threitol units can effect enantioselective epoxidation of (E)-β-methylstyrene in up to 70% ee. For the asymmetric styrene oxidation, a lower enantioselectivity of 40% ee was obtained (c.f. 62% ee, see Table 6.3) when

Table 6.3 Epoxidation of some aromatic alkenes by $[Ru^{VI}(L^1)O_2]$ (1a).

entry	alkenes	solvent	epoxide yield (%)	% ee (abs. config.)
1	Ph	C_6H_6	64[c]	62 (R)
		C_6H_6	62	40 (R)[d]
		CH_2Cl_2	39	41 (R)
		MeCN	13	33 (R)
2	Cl (aromatic vinyl)	C_6H_6	75	60 (R)
3	Ph ... Me	C_6H_6	90 (>99 % trans)	67 (1R,2R)
		C_6H_6(0°C)	90 (>99 % trans)	70 (1R,2R)
		CH_2Cl_2	58 (>99 % trans)	32 (1R,2R)
		EtOAc	82 (>99 % trans)	38 (1R,2R)
4	Ph ... Cl	C_6H_6	70	76 (1R,2R)
5	Ph Me	C_6H_6	75 (> 99% cis)	40 (1R,2S)
		CH_2Cl_2	68 (95% cis, 5% trans)	18 (1R,2S)
6	(naphthalene ring)	C_6H_6	88	20 (1R,2S)

the reaction was carried out without pyrazole. It is believed that the pyrazole could prevent the Ru(IV) intermediate from undergoing further reduction to a Ru(II) porphyrin by forming the $[Ru^{IV}(L^1)(pz)_2]$ complex, where the Ru(II) species is known to racemize chiral epoxides[8]. The asymmetric (E)-β-methyl-styrene epoxidation by $[Ru^{VI}(L^1)O_2]$ exhibits remarkable solvent dependence. Benzene is the solvent of choice, and the use of polar solvents such as dichloromethane or ethyl acetate would lead to lower enantioselectivities of 32 and 38% ee, respectively. Other dioxoruthenium derivatives bearing gem-diethyl, $[Ru^{VI}(L^2)O_2]$, and gem-cyclopentyl groups, $[Ru^{VI}(L^3)O_2]$, at the threitol units afforded a lower ee of 60% and 55% ee for the styrene oxidation.

Table 6.3 depicts the results of the asymmetric epoxidation of some aromatic alkenes.

REFERENCES

1. Jacobsen, E.N., Zhang, W., Muci, A.R., Ecker, J.R., Deng, L. *J. Am. Chem. Soc.*, 1991, **113**, 7063.
2. Wang, Z.-X., Tu, Y., Frohn, M., Zhang, J.-R., Shi, Y. *J. Am. Chem. Soc.*, 1997, **119**, 11224.
3. Zhang, W., Jacobsen, E.N. *J. Org. Chem.*, 1991, **56**, 2296.
4. Deng, L., Jacobsen, E.N. *J. Org. Chem.*, 1992, **57**, 4320.
5. Bennani, Y.L., Hanessian, S. *Chemical Reviews*, 1997, **1997**, 3161.

6. Mio, S., Kumagawa, Y., Sugai, S. *Tetrahedron*, 1991, **47**, 2133.
7. Zhang, R., Yu, W.-Y., Lai, T.-S., Che, C.-M. *Chem. Commun.*, 1999, 409.
8. Groves, J.T., Ahn, K.T., Quinn, R. *J. Am. Chem. Soc.*, 1988, **110**, 4217.

7 Asymmetric Hydroxylation and Aminohydroxylation

CONTENTS

7.1 ASYMMETRIC AMINOHYDROXYLATION OF 4-METHOXYSTYRENE

P. O'BRIEN, S.A. OSBORNE and D.D. PARKER

Department of Chemistry, University of York, Heslington, York YO10 5DD, UK

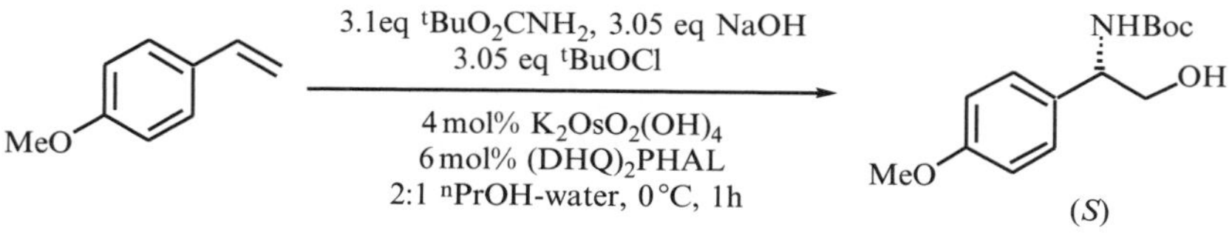

Materials and equipment

- *tert*-Butyl carbamate, 545 mg, 4.65 mmol
- *n*-Propanol, 24.2 mL
- Sodium hydroxide, 183 mg, 4.6 mmol
- Water, 12.2 mL
- *tert*-Butylhypochlorite, freshly prepared, 0.53 mL, 4.6 mmol
  tert-Butyl hypochlorite was freshly prepared according to the literature procedure[1,2] using sodium hypochlorite (5% chlorine, purchased from Acros Organics, catalogue number 41955). The quality of the sodium

hypochlorite is important for the preparation of good quality *tert*-butyl hypochlorite. *tert*-Butyl hypochlorite can be stored in a foil-covered flask in the freezer for up to 3 weeks without any noticeable change in performance.

- DHQ$_2$PHAL, 71 mg, 0.09 mmol
- 4-Methoxystyrene, 201 mg, 1.5 mmol
- Potassium osmate dihydrate [K$_2$OsO$_2$(OH)$_4$], 22.5 mg, 0.06 mmol
- Saturated aqueous sodium sulfite solution
- Ethyl acetate, petroleum ether (40–60 °C)
- Brine
- Magnesium sulfate

- 50 mL round-bottomed flask with magnetic stirrer bar
- Magnetic stirrer
- Separating funnel, 100 mL
- Rotary evaporator
- Flash chromatography column, 3 cm diameter

Procedure

1. In a 50 mL round-bottomed flask equipped with a magnetic stirrer bar were placed *tert*-butyl carbamate (545 mg) and *n*-propanol (6 mL). A solution of sodium hydroxide (183 mg) in water (12.2 mL) and *tert*-butyl hypochlorite (0.53 mL) were added to the solution. The resulting solution was stirred for 5 minutes and cooled to 0 °C. Then a solution of DHQ$_2$PHAL (71 mg) in *n*-propanol (6 mL), a solution of 4-methoxystyrene in *n*-propanol (12.2 mL) and potassium osmate dihydrate (22.5 mg) were added sequentially to give a green solution. After 1 hour at 0 °C, the reaction mixture had turned from green to yellow.
2. The reaction was monitored by TLC (eluent: petroleum ether–ethyl acetate, 1:1). Visualized by ninhydrin dip, the product stained brown-orange, R_f 0.43. The 4-methoxystyrene (visualized by UV) has R_f 0.72.
3. After completion of the reaction, saturated aqueous sodium sulfite solution (10 mL) was added and the mixture stirred for 15 minutes. Ethyl acetate (5 mL) was added and the two phases were separated. The aqueous layer was extracted with ethyl acetate (3 × 5 mL). The combined organic extracts were washed with brine (20 mL), dried over magnesium sulfate, filtered and concentrated using a rotary evaporator to give the crude product.
4. Purification by flash column chromatography on silica (eluent: petroleum ether–ethyl acetate, 2:1) gave a crystalline solid (*S*)-*N*-(*tert*-butoxycarbonyl)-1-(4-methoxyphenyl)-2-hydroxyethylamine (296 mg, 74 %).

   The ee (98 %) was determined by HPLC (Chiralcel OD-H column, flow 1 mL/min, eluent: heptane–*iso*-propanol, 95:5); (*S*)-enantiomer: R_t 11.7 min, (*R*)-enantiomer: R_t 10.3 min.

^{1}H NMR (270 MHz, CDCl$_3$) 7.26–7.19 (m, 2 H, Ar); 6.90–6.87 (m, 2 H, Ar); 5.16 (br s, 1 H, NH); 4.73 (br s, 1 H, ArCHN); 3.82 (br s, 2 H, CH$_2$O); 3.80 (s, 3 H, MeO); 1.43 (s, 9 H, CMe$_3$).
IR (CHCl$_3$, cm^{-1} 3611 (O–H); 3442 (N–H); 1707 (C=O)).
mp 139–141 °C (from petroleum ether-ethyl acetate, 2:1).
$[\alpha]_D$ + 62.6 (c 1.0 in EtOH).

7.1.1   CONCLUSION

The asymmetric aminohydroxylation[3] of 4-methoxystyrene using DHQ$_2$ PHAL as the ligand actually produces an 85:15 mixture of (S)-N-(tert-butoxycarbonyl)-1-(4-methoxyphenyl)-2-hydroxyethylamine and its regioisomer as shown by ^{1}H NMR spectroscopy on the crude product mixture. The regioisomer is lower running by TLC. The product is separated from the regiosiomer and from excess tert-butyl carbamate by careful flash column chromatography: this is a limitation of the methodology.

(R)-N-(tert-Butoxycarbonyl)-1-(4-methoxyphenyl)-2-hydroxyethylamine(ee, 96%) can be prepared using DHQD$_2$PHAL as the ligand but this results in the production of more of the unwanted regioisomer: a 68:32 mixture of the (R)-product and its regioisomer were obtained. This gives a lower isolated yield of (R)-N-(tert-butoxycarbonyl)-1-(4-methoxyphenyl)-2-hydroxyethylamine (65%) as compared to its enantiomer (74%). The same trend is observed with other styrene derivatives[2]. A wide range of styrene derivatives give high enantiomeric excesses using these conditions.[2,4]

7.2   ASYMMETRIC DIHYDROXYLATION OF (1-CYCLOHEXENYL) ACETONITRILE

JEAN-MICHEL VATÈLE

Université Claude Bernard Laboratoire de Chimie Organique 1, CPE-Bât. 308, 43, Boulevard du 11 November 1918, 69622 Villeurbanne Cedex, France.

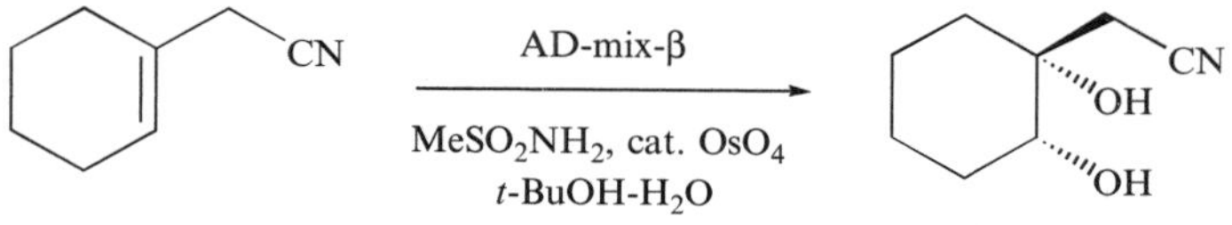

(1R, 2R)

Materials and equipment

- tert-Butanol, 70 mL
- Water, 70 mL

- AD-mix-β, 20 g
- Methanesulfonamide, 1.36 g
- Osmium tetroxide (4 wt% solution in water), 0.36 mL
- (1-Cyclohexenyl)acetonitrile, 1.73 g
- Sodium metabisulfite ($Na_2S_2O_5$), 14 g
- Dichloromethane, 360 mL
- Magnesium sulfate
- Ether, petroleum ether
- (40–63 μm) Silica gel 60, 30 g

- 250 mL Round-bottomed flask with a magnetic stirrer bar
- Magnetic stirrer
- Separating funnel, 500 mL
- Sintered glass funnel (4 cm)
- Flash column chromatography (30 cm ×2.5 cm)

Procedure

1. In a 250 mL round-bottomed flask equipped with a magnetic stirrer bar were placed a 1:1 mixture of *tert*-butanol and water (140 mL), AD-mix-β (20 g) and methanesulfonamide (1.36 g)[5].

   100 g of AD-mix-β are made up of potassium osmate (0.052 g), $(DHQD)_2PHAL$ (0.55 g), $K_3Fe(CN)_6$ (70 g), K_2CO_3 (29.4 g).
2. The mixture was stirred for a few minutes at room temperature until two clear phases were produced. To the ice-chilled reaction mixture were successively added osmium tetroxide (4 wt% in water, 0.36 mL) and (1-cyclohexenyl)-acetonitrile (1.73 g). The reaction mixture was stirred vigorously for 8 hours at 0 °C.
3. The reaction was monitored by TLC (eluent: petroleum ether–ether, 4:1). Olefin and diol spots, visualized by iodine vapour, have R_f values of 0.43 and zero respectively.
4. Sodium metabisulfite (14 g) was added to the reaction mixture and stirring was continued for 1 hour.
5. The reaction mixture was transferred to a separating funnel and extracted three times with dichloromethane (3 ×120 mL). The combined organic phases were dried over magnesium sulfate, filtered and concentrated using a rotary evaporator. The residue was purified by flash chromatography on silica gel (eluent: ether–petroleum ether, 3:2) to give a crystalline product (2.1 g, 94 %), mp 95–101 °C, $[\alpha]^{20}_D - 1.6$ (c 1, CHCl$_3$)[6].

   The ee (66–71 %) was determined on the acetonide derivative by GC analysis.

   ^{1}H NMR (200 MHz, CDCl$_3$: δ 1.3–1.8 (m, 7 H), 2.0 (m, 1 H), 2.58 (d + brs, 2 H, J 17 Hz, C$\underline{H}_a$CN, OH), 2.73 (d + brs, 2 H, J 17 Hz, C$\underline{H}_b$CN, OH), 3.52 (dd, 1 H, J 4.4 and 10.3 Hz, C$\underline{H}$OH).

   ^{13}C NMR (50.3 MHz, CDCl$_3$): δ 20.6, 23.7, 28.8, 30.4, 34.6, 72.1, 117.9.

7.2.1   (R,R)-(1,2-DIHYDROXYCYCLOHEXYL)ACETONITRILE ACETONIDE

Materials and equipment

- Tetrahydrofuran, 40 mL
- (1,2-Dihydroxycyclohexyl)acetonitrile (66 % ee), 2.1 g
- Camphorsulfonic acid (CSA), 0.04 g
- 2-Methoxypropene, 2.5 mL
- Ether, petroleum ether
- (40–63 μm) Silica gel 60, 40 g
- Hexane

- 100 mL Round-bottomed flask with a magnetic stirrer bar
- Magnetic stirrer plate
- Flash column chromatography (30 cm × 2.5 cm)
- Rotary evaporator

Procedure

1. (1,2-Dihydroxycyclohexyl)acetonitrile (2.1 g) was placed in a 100 mL round-bottomed flask equipped with a magnetic stirrer bar. Dry tetrahydrofuran (40 mL) was added followed by CSA (0.04 g) and 2-methoxypropene (2.5 mL). The solution was stirred at room temperature for 90 minutes.
2. The reaction was monitored by TLC (eluent: petroleum ether–ether, 1:2). Diol and acetonide spots, visualized by p-anisaldehyde dip, have R_f values of 0 and 0.35 respectively.
3. The reaction mixture was concentrated using a rotary evaporator. The residue was chromatographed on silica gel (eluent: ether–petroleum ether, 1:3) to afford a white solid (97 %).
4. The solid was crystallised twice in hexane to afford a compound (1.68 g) with a high enantiomeric purity (94.7 % ee), mp 55–60 °C, $[\alpha]_D^{20}$ −27.6 (c 0.7, CHCl$_3$).

   The enantiomeric excess was determined by capillary GC analysis (Lipodex E.MN, 25 m, 0.25 mm ID, temperature column 80–150 °C, 5 °C/min). (S,S)-enantiomer: R_t 19.25 min, (R,R)-enantiomer: R_t 19.69 min.

   The absolute configuration was determined by chemical correlation with the known (2R)-2-allyl-2-hydroxycyclohexanone.

^{1}H NMR (200 MHz, CDCl$_3$): δ 1.32 (s, 3 H, CH$_3$), 1.5 (s, 3 H, CH$_3$), 1.5–1.7 (m, 7 H), 2.1–2.2 (m, 1H), 2.58 (d, 1 H, J 17 Hz, C$\underline{H}_a$CN), 2.67 (d, 1 H, J 17 Hz, C$\underline{H}_b$CN), 4.1 (brs, 1 H, C$\underline{H}$OCMe$_2$).
^{13}C NMR (50.3 MHz, CDCl$_3$): δ 19.5, 22.6, 25.5, 26.2, 26.8, 28.3, 34.6, 76.0, 77.2, 108.3, 116.9.

7.2.2 CONCLUSION

Osmium-catalysed asymmetric dihydroxylation allowed an efficient transformation of (1-cyclohexenyl)acetonitrile to enantioenriched (1,2-isopropylidenedioxy)cyclohexylacetonitrile in good yield (65%) and high enantiomeric purity (94.7% ee). This acetonide can be easily transformed in a few steps to α-ketols bearing allyl or 3-trimethylsilylpropargyl groups, precursors of neurotoxic alkaloids histrionicotoxins.[7, 8] This methodology enabling the preparation of substituted α-ketols is superior to that described in the literature.[8, 9] This method which involves a chromatographic resolution of diastereomeric ketals presents several drawbacks such as a low efficiency of the chromatographic separation ($\Delta R_f = 0.05$) and occurrence of a partial racemization during acid deketalization.

Sharpless asymmetric dihydroxylation has been successfully applied to (1-cyclopentenyl)acetonitrile. Using (DHQ)PHN as a ligand in place of (DHQ)$_2$PHAL,* one of the components of AD-mix-α, (S,S)-(1,2-dihydroxycyclopentyl)acetonitrile was obtained after two recrystallizations, in 50% yield and 90% ee.

REFERENCES

1. Mintz, M.J., and Walling, C. *Org. Synth.*, 1983, Coll. Vol. V, 183.
2. O'Brien, P., Osborne, S.A., and Parker, D.D. *J. Chem. Soc., Perkin Trans. 1*, 1998, 2519; O'Brien, P., Osborne, S.A., Parker, and D.D. *Tetrahedron Lett.*, 1998, **39**, 4099.
3. O'Brien, P. *Angew. Chem. Int. Ed. Engl.*, 1999, **38**, 326.
4. Reddy, K.L., and Sharpless, K.B., *J. Am. Chem. Soc.*, 1998, **120**, 1207.
5. For the use of asymmetric dihydroxylation and its modifications in organic synthesis see: Kolb, H.C., Van Nieuwenhze, M.S. and Sharpless, K.B. *Chem. Rev.*, 1994, **59**, 2483–547.
6. Devaux, J.M., Goré, J. and Vatèle, J.M. *Tetrahedron: Asymmetry*, 1998, **9**, 1619–26.
7. Compain, P., Goré, J. and Vatèle, J.M. *Tetrahedron Lett.*, 1995, **36**, 4063–4.
8. Compain, P., Goré, J. and Vatèle, J.M. *Tetrahedron Lett.*, 1996, **52**, 6647–64.
9. Compain, P., Goré, J. and Vatèle, J.M. *Tetrahedron Lett.*, 1995, **36**, 4059–62.

---

* (DHQ)PHN and (DHQ)$_2$PHAL are the respective abbreviations of dihydroquinine 9-phenanthryl ether and of dihydroquinidine 1,4-phthalazinediyl diether.

8   Asymmetric Sulfoxidation

CONTENTS

8.1   ASYMMETRIC OXIDATION OF SULFIDES AND KINETIC RESOLUTION OF SULFOXIDES

LAURA PALOMBI and ARRIGO SCETTRI

Dipartimento di Chimica, Universita di Salerno, 84081 Baronissi (Salerno), Italy

8.1.1   ASYMMETRIC OXIDATION OF 4-BROMOTHIOANISOLE

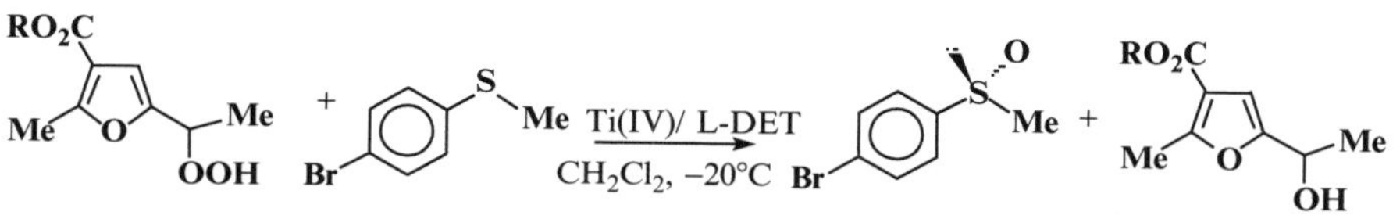

1   2   (R)-3   4

R: Cholestanyl

Materials and equipment

- Anhydrous dichloromethane (99.8 %), 3.5 mL
- 4-Bromothioanisole (97 %), 61 mg, 0.3 mmol
- Titanium(IV) isopropoxide (97 %), 0.09 mL, 0.3 mmol
- Diethyl L-tartrate (99 %), 0.21 mL, 1.2 mmol
- 5-(1-Hydroperoxyethyl)-2-methyl-3-furoic acid 5α-cholestan-3β-yl ester (**1**)[1c] (mixture of diastereo-isomers), 171 mg, 0.3 mmol
- Ethyl acetate, *n*-hexane
- Anhydrous sodium sulfate

- Silica gel 60 (230–400 mesh ASTM)

- One 10 mL round-bottomed flask with magnetic stirrer bar
- Magnetic stirrer
- Refrigerator bath (−22 °C)
- Büchner funnel (6 cm)
- Büchner flask
- Filter paper (589 Blue ribbon)
- Rotary evaporator
- Chromatography column (15 mm diameter)

Procedure

1. In a 10 mL round-bottomed flask equipped with a magnetic stirrer bar were placed, under an argon atmosphere, anhydrous dichloromethane (2 mL) and diethyl L-tartrate (0.21 mL).
2. The mixture was cooled to −22 °C, then titanium(IV) isopropoxide (0.09 mL) and sulfide (61 mg solved in 1.5 mL of anhydrous dichloromethane) were added. Stirring was maintained for 20 minutes and hydroperoxide (171 mg) added to the mixture.
3. The reaction was then continued for 2 hours until completion [monitoring by TLC (eluent: *n*-hexane–ethyl acetate, 5:1. Detector: UV lamp at 254 nm) indicated complete consumption of the hydroperoxide].
4. After completion, water (1.5 mL) was added to the mixture and vigorous stirring continued for 2 hours at room temperature. The resulting white gel was diluted with ethyl acetate and filtered over a filter paper in a Büchner funnel. The solution was dried over sodium sulfate, filtered and concentrated using a rotary evaporator.
5. Purification of the crude mixture was performed by flash chromatography to afford pure 5-(1-hydroxyethyl)-2-methyl-3-furoic acid 5α-cholestan-3β-yl ester (**4**) (90 % yield) and (*R*)-4-bromophenyl methyl sulfoxide (**3**) (61 % yield).
   The ee (92 %) was determined on representative sample by ^{1}H-NMR analysis in the presence of *R*-(−)-(3,5-dinitrobenzoyl)-α-methylbenzyl amine.
   ^{1}H-NMR (200 MHz, CDCl$_3$): 7.68 (d, 2H, J = 8.6 Hz); 7.52 (d, 2H, J = 8.6 Hz); 2.72 (s, 3H).

Table 8.1 Asymmetric oxidation of methylsulfides (**R–S–Me**) with **1**.

Entry	R	Reac. time (h)	Yield (%)	e.e. (%)
a	4-NO$_2$-C$_6$H$_4$	3	58	>95 (R)
b	2-Br-C$_6$H$_4$	7	58	94 (R)
c	4-Cl-C$_6$H$_4$	2	63	90 (R)
d	4-Me-C$_6$H$_4$	2	74	84 (R)
e	*n*-octyl	2	78	39 (R)

Conclusion

The use of furylhydroperoxides[1] has facilitated an operationally simple procedure, alternative to the one reported by Kagan[2]. Oxidation takes place rapidly and very high e.e.s have been obtained, especially in the case of aryl methyl sulfides, while overoxidation to sulfone can be reduced to a great extent (<3%) under the proposed experimental conditions.

Modified catalytic procedures involving the employment of different chiral ligands, such as binaphthol[3], 2,2,5,5-tetramethyl-3,4-hexanediol[4], 1,2-diarylethane-1,2-diols[5] are often less selective and very variable values of e.e.s can be observed.

8.1.2   KINETIC RESOLUTION OF RACEMIC 4-BROMOPHENYL METHYL SULFOXIDE

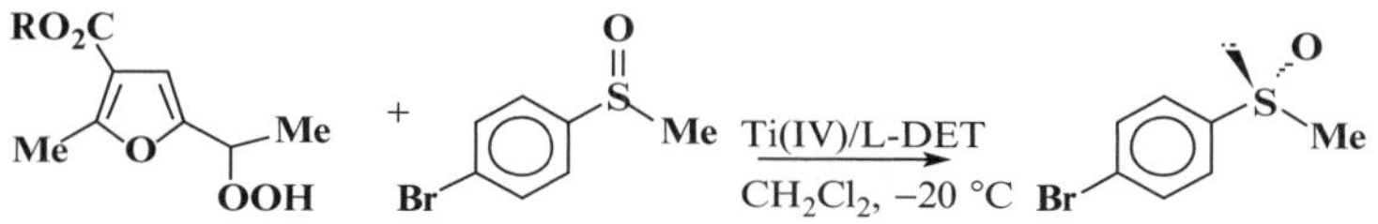

R: Cholestanyl

+sulfone

Materials and equipment

- Anhydrous dichloromethane (99.8%), 3.5 mL
- Racemic 4-bromophenyl methyl sulfoxide (97%), 110 mg, 0.5 mmol
- Titanium(IV) isopropoxide (97%), 0.075 mL, 0.25 mmol
- Diethyl L-tartrate (99%), 0.175 mL, 1 mmol
- 5-(1-Hydroperoxyethyl)-2-methyl-3-furoic acid 5α-cholestan-3β-yl ester (1)[1c] (mixture of diastereoisomers), 228 mg, 0.4 mmol
- Ethyl acetate, n-hexane
- Anhydrous sodium sulfate
- Silica gel 60 (230–400 mesh ASTM)

- One 10 mL round-bottomed flask with magnetic stirrer bar
- Magnetic stirrer
- Refrigerator Bath (−22 °C)
- Büchner funnel (6 cm)
- Büchner flask
- Filter paper (589 Blue ribbon)
- Rotary evaporator
- Chromatography column (15 mm diameter)

Procedure

1. In a 10 mL round-bottomed flask equipped with a magnetic stirrer bar were placed, under an argon atmosphere, anhydrous dichloromethane (2 mL) and diethyl L-tartrate (0.21 mL).
2. The mixture was cooled to −22 °C, then titanium(IV) isopropoxide (0.09 mL) and hydroperoxide (228 mg) were added. Stirring was maintained for 10 minutes and racemic sulfoxide (110 mg solved in 1.5 mL of anhydrous dichloromethane) added to the mixture.
3. The reaction was continued for 13 hours until completion [monitoring by TLC (eluent: n-hexane–ethyl acetate, 5:1. Detector: UV lamp at 254 nm) indicated complete consumption of the hydroperoxide].
4. After completion, water (1.5 mL) was added to the mixture and vigorous stirring continued for 2 hours at room temperature. The resulting white gel was diluted with ethyl acetate and filtered over a filter paper in a Büchner funnel. The solution was dried over sodium sulfate, filtered and concentrated using a rotary evaporator.
5. Purification of the crude mixture was performed by flash chromatography to afford pure 5-(1-hydroxyethyl)-2-methyl-3-furoic acid 5α-cholestan-3β-yl ester (86 % yield), 4-bromophenyl methyl sulfone (59 % yield) and (R)-4-bromophenyl methyl sulfoxide (34 % yield).

   The ee (>95 %) was determined on representative sample by ¹H-NMR analysis in the presence of R-(−)-(3,5-dinitrobenzoyl)-α-methylbenzyl amine.

   Stereoselection factors have been determined according to Kagan's equation[6].

Table 8.2   Kinetic resolution of racemic sulfoxides (R–SO–Me) with **1**.

Entry	R	Reac. Time (h)	Yield (%)	e.e. (%)	E
a	n-octyl	13	31	94	7.8
b	4-ClC$_6$H$_4$	14	29	> 95	> 7.4
c	C$_6$H$_5$	23	32	82	5.3
d	4-MeC$_6$H$_4$	20	42	83	10
e	4-NO$_2$C$_6$H$_4$	14	40	> 95	> 15.7

Conclusion

The above procedure can be exploited for the asymmetric oxidation of racemic sulfoxide[1], and high stereoselection can be frequently observed. Moreover unreacted R-sulfoxides were always recovered as the most abundant enantiomers, kinetic resolution and asymmetric oxidation being two enantioconvergent processes. Thus, by the combined routes, higher enantioselectivity can be observed with dialkyl sulfoxides, usually obtained with poor to moderate e.e.s.

Furylhydroperoxides of type **1** or cumyl hydroperoxide can be used according to the particular sulfoxide to be resolved. Other procedures, involving different chiral auxiliaries[3,4,7] lead to much lower stereoselection factors (≈ 3).

REFERENCES

1. (a) Lattanzi, A., Bonadies, F., Scettri, A. *Tetrahedron: Asymmetry*, 1997, **8**, 2141–51. (b) Lattanzi, A., Bonadies, F., Scettri, A. *Tetrahedron: Asymmetry*, 1997, **8**, 2473–78.(c) Palombi, L., Bonadies, F., Pazienza, A., Scettri A. *Tetrahedron: Asymmetry*, 1998, **9**, 1817–22.
2. (a) Zhao, S.H., Samuel, O., Kagan, H.B. *Tetrahedron*, 1987, **43**, 5135–44. (b) Brunel, J.M., Kagan, H.B. *Synlett.*, 1996, 404–5.
3. Komatsu, N., Hashizume, M., Sugita, T., Uemura, S. *J. Org. Chem.*, 1993, **58**, 4529–33.
4. Yamanoi, Y., Imamoto, T.*J. Org. Chem.*, 1997, **62**, 8560–4.
5. Superchi, S., Rosini, C. *Tetrahedron: Asymmetry*, 1997, **8**, 349–52.
6. Kagan, H.B., Fiaud, J.C. *Top. Stereochem.*, 1988, **18**, 249.
7. Noda, K., Hosoya, N., Irie, R., Yamashita, Y., Katsuki, T. *Tetrahedron*, 1994, **32**, 9609–18.

9 Asymmetric Reduction of Ketones Using Organometallic Catalysts

CONTENTS

9.1 INTRODUCTION

Asymmetric catalytic reduction reactions represent one of the most efficient and convenient methods to prepare a wide range of enantiomerically pure compounds (i.e. α-amino acids can be prepared from α-enamides, alcohols from ketones and amines from oximes or imines). The chirality transfer can be accomplished by different types of chiral catalysts: metallic catalysts are very efficient for the hydrogenation of olefins, some ketones and oximes, while nonmetallic catalysts provide a complementary method for ketone and oxime hydrogenation.

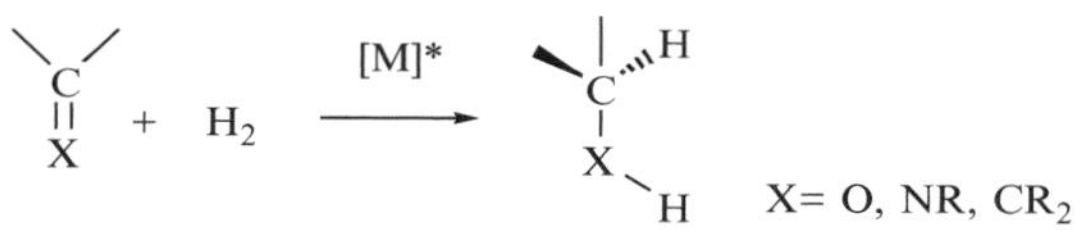

X = O, NR, CR$_2$

Enantioselective catalysis using chiral metal complexes provides a flexible method for asymmetric hydrogenation. The metallic elements possess a variety of catalytic activities and their combination with organic ligands or auxiliaries that direct the steric course can give very efficient catalytic complexes. Well-designed chiral metal complexes can discriminate precisely between enantiotopic groups, or faces, and catalyse the formation of a wide range of natural and unnatural substances with high enantiomeric purity. Asymmetric reduction with a transition metal can use molecular hydrogen (hydrogen gas) as the source of hydrogen or nonhazardous organic molecules as donors of hydrogen such as formic acid or 2-propanol. This last method, hydrogen transfer, can provide a complement to the catalytic reduction using molecular hydrogen[1].

The first enantioselective hydrogenation of unsaturated compounds was using metallic catalysts deposited on chiral supports in the 1930s[2]. In the 1950s, using this method an enantioselectivity exceeding 60 % was obtained[3]. Knowles[4] and Horner[5] in 1968 reported homogeneous asymmetric hydrogenation using rhodium–chiral tertiary phosphine complexes.

Nonmetallic systems (Chapter 11) are efficient for catalytic reduction and are complementary to the metallic catalytic methods. For example lithium aluminium hydride, sodium borohydride and borane–tetrahydrofuran have been modified with enantiomerically pure ligands[6]. Among those catalysts, the chirally modified boron complexes have received increased interest. Several ligands, such as amino alcohols[7], phosphino alcohols[8,9] and hydroxysulfoximines[10], complexed with the borane, have been found to be selective reducing agents.

In 1969, Fiaud and Kagan[11] tested ephedrine boranes but achieved only 3.6–5 % enantiomeric excess in the reduction of acetophenone. Itsuno et al.[12] reported in 1981 an interesting enantioselective reduction of a ketone using an amino alcohol–borane complex as a catalyst. Buono[13] investigated and developed the reactivity of phosphorus compounds as ligands in borane complexes for asymmetric hydrogenation.

Enzyme reductions of carbonyl groups have important applications in the synthesis of chiral compounds (as described in Chapter 10). Dehydrogenases are enzymes that catalyse, for example, the reduction of carbonyl groups; they require co-factors as their co-substrates. Dehydrogenase-catalysed transformations on a practical scale can be performed with purified enzymes or with whole cells, which avoid the use of added expensive co-factors. Bakers' yeast is the whole cell system most often used for the reduction of aldehydes and ketones. Biocatalytic activity can also be used to reduce carbon–carbon double bonds. Since the enzymes for this reduction are not commercially available, the majority of these experiments were performed with bakers' yeast[14].

In summary, the asymmetric hydrogenation of olefins or functionalized ketones catalysed by chiral transition metal complexes is one of the most practical methods for preparing optically active organic compounds. Ruthenium- and rhodium–diphosphine complexes, using molecular hydrogen or hydrogen transfer, are the most common catalysts in this area. The hydrogenation of simple ketones has proved to be difficult with metallic catalysts. However, asymmetric borane complexes are used as nonmetallic catalysts for the hydrogenation of simple ketones, like acetophenone. They can give the corresponding alcohol in high yield and enantiomeric excess. The use of oxidoreductases contained in bakers' yeast can give good results for reduction of carbonyl and carbon–carbon double bonds.

In contrast to the enantioselective reduction of alkenes or ketones, few catalytic systems have been described for the enantioselective reduction of imines. The nature of the N-substituent and the E/Z-isomerism caused by the carbon–nitrogen double bond of the substrate are important parameters for the control of the enantioselectivity[15]. The enantioselective hydrogenation of carbon–nitrogen bonds has been reported to occur in the presence of iridium, ruthenium,[1,16,17] titanium,[18–20] zirconium or cobalt[21], with chiral diphosphine ligands. Iridium catalysts have been successfully employed in the hydrogenation of N-arylimines.[22–25] Rhodium catalysts have been used for the reduction of imines and nitrones.[26,27] The employment of chiral auxiliaries for the activation of borane reagents in carbon–nitrogen double bond reduction has been shown to induce high enantioselectivities. However, it is quite difficult to maintain the high enantiomeric excesses, obtained by using stoichiometric amounts of such auxiliaries, in a catalytic version of the reduction[28]. The substrates which lead to the best results are oxime ethers. The oxazaborolidine derived from valinol is a very efficient chiral auxiliary for oxime ether reduction with borane. The oxazaphospholidine derived from prolinol (developed by Buono et al.) induce enantioselectivity in the reduction of imines[8].

In this chapter and in Chapters 10–12, we will review and validate some methods for asymmetric (transfer) hydrogenation of carbon–oxygen and carbon–carbon double bonds catalysed by non-metallic systems, homogeneous transition metal catalysts and biocatalysts. Reduction of carbon–nitrogen double bond systems will be reported in another volume of this series.

9.2   ASYMMETRIC HYDROGENATION USING A METAL CATALYST: [RU((S)-BiNAP)]

Few catalysts have been found to produce chiral alcohols from a range of ketones with both high levels of absolute stereocontrol and high catalytic efficiencies[29]. Metallic catalysts are generally used for the asymmetric hydrogenation of functionalized ketones. High enantioselectivities (>98% ee) have been observed in the hydrogenation of a variety of β-keto esters using 2,2'-bis(phosphino)-1,1'-biaryl (BINAP)-derived catalysts such as ruthenium–BINAP[30]. For the hydrogenation of simple ketones, nonmetallic catalysts like chiral borane complexes are often preferred (Chapter 11).[31,32] Biocatalytic reduction with bakers' yeast is used for reduction of ketones such as β-ketoesters, β-diketones, α-hydroxy ketones, aliphatic and aromatic ketones[33] (Chapter 10).

The mechanism of a metal-catalysed reduction is believed to proceed as described in Figure 9.1.

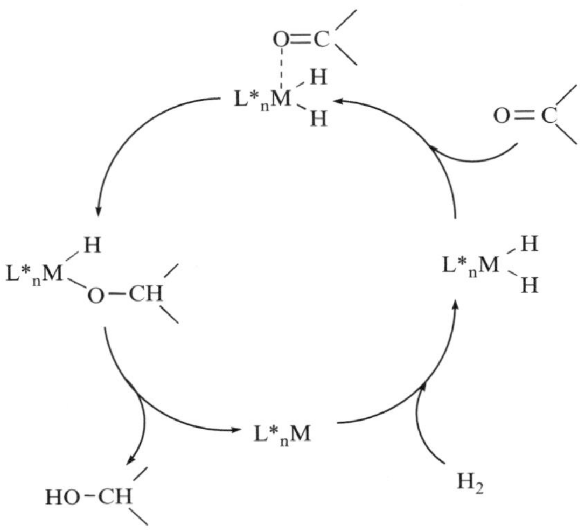

Figure 9.1 Catalytic cycle for hydrogenation of carbonyl compounds.

The method developed by Noyori using [Ru(BiNAP)Cl$_2$](NEt$_3$) as catalyst requires high temperatures and/or hydrogen pressures (100 atm)[30]. As this method needs a specific set-up for hydrogenation at high pressure, it was not possible for us to validate this procedure. However, Genêt reported[34, 35] that the catalyst [Ru(BiNAP)Br$_2$](acetone) can give good results for the hydrogenation of β-ketoesters at atmospheric pressure.

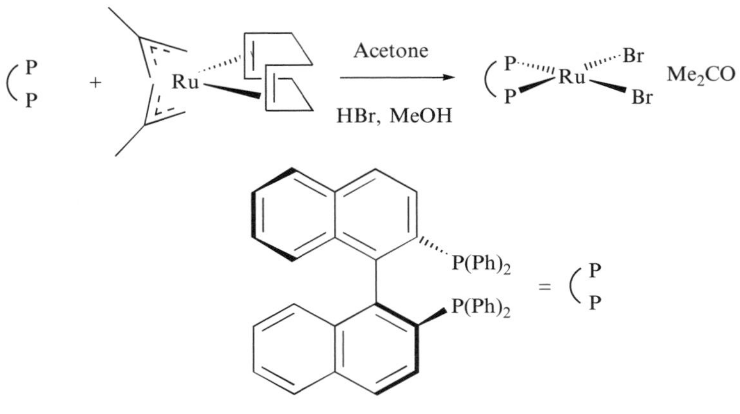

S-BiNAP

Materials and equipment

- [Ru(allyl)$_2$(COD)$_n$], 6 mg, 0.019 mmol, 0.02 eq
- (S)-(−)-2,2'-Bis(diphenylphosphino)-1,1'-binaphthyl [(S)-BiNAP], 21 mg, 0.033 mmol, 0.035 eq
- Anhydrous acetone degassed (nitrogen) for 30 minutes, 30 mL
- Anhydrous methanol degassed (nitrogen) for 30 minutes, 30 mL
    As the catalyst is very sensitive to oxygen, the solvents were degassed with nitrogen just before use.
- Hydrobromic acid 48 %, 1.0 mL
- Hydrobromic acid solution 0.6 M prepared by mixing 9 mL of degassed methanol and 1 mL of hydrobromic acid 48 %
- Methyl acetoacetate, 100 µL, 0.93 mmol
    Methyl acetoacetate was dried with magnesium sulfate which was activated at 500 °C for 2 hours cooled under vacuum and stored under nitrogen.
- Petroleum ether, ethyl acetate, methanol
- Silica gel 60 (0.063–0.04 mm)
- p-Anisaldehyde

- 50 mL Schlenk tube
- Egg-shape magnetic stirrer bar
- Low-pressure hydrogenation apparatus fitted with a gas burette system to measure the hydrogen consumed[36]
- Rotary evaporator
- Kugelrohr apparatus

Procedure

1. A 50 mL Schlenk tube was dried overnight in a oven at 150 °C, cooled under vacuum and flushed with nitrogen.
2. The Schlenk tube was filled with [Ru(allyl)$_2$(COD)$_n$] (6 mg), ((S)-BiNAP) (21 mg) and purged twice using vacuum/nitrogen cycles. Anhydrous acetone was added (2 mL) to give a white suspension. The solution was stirred for 30 minutes at room temperature.
3. To this suspension was added a solution of HBr (0.6 M, 0.11 mL). The suspension was stirred for 30 minutes at room temperature. After 15 minutes a yellow precipitate appeared.
4. The solvent was removed by evaporation under high vacuum over 3 hours to give a yellow powder, which was used as the catalyst without any further purification.

5. The Schlenk tube containing the catalyst was filled with degassed methanol (20 mL) and methyl acetoacetate (100 µL); the brown suspension was placed under nitrogen.

6. The Schlenk tube was connected to a low-pressure hydrogenation apparatus fitted with a gas burette system to measure the hydrogen consumed. The Schlenk tube was flushed through three cycles (reduced pressure/hydrogen) and then placed under an atmospheric pressure of hydrogen. The burette was filled with 200 mL of hydrogen.

Never allow naked flames in the vicinity when hydrogen is being used. Avoid the formation of air–hydrogen mixtures. Any electrical apparatus in the vicinity must be spark-proof. It is far better for the apparatus to be kept in a separate room specially designed for hydrogenations.

7. The solution was vigorously stirred to increase the contact area of the reactants with the hydrogen atmosphere. The solution became lighter brown. The reaction was stopped after 48 hours (no more hydrogen was consumed), by removing the hydrogen using reduced pressure.

8. The reaction was monitored by TLC (eluent: petroleum ether–ethyl acetate; 75:25). The methyl acetoacetate was UV active, stained yellow with *p*-anisaldehyde, R_f 0.5. No starting material remained after 48 hours.

9. The solution was filtered through a pad of silica gel to remove the catalyst and the filter residue was washed with methanol. The solvent was removed under reduced pressure using a rotary evaporator (water bath at 30 °C) to give a slightly brown oil.

10. The residue was distilled with a Kugelrohr apparatus under water aspirator vacuum (approximately 20 mbar, 140 °C) to give (*S*)-methyl-(3)-hydroxybutanoate (105 mg, 99%).

The ee (>98%) was determined by chiral GC (Lipodex® E, 25 m, 0.25 mm ID, temperatures: column 90 °C isotherm, injector 250 °C, detector 250 °C, mobile phase helium). R_t (*R*)-enantiomer: 13.5 min; R_t (*S*)-enantiomer: 15.2 min.

^{1}H NMR (200 MHz, CDCl₃): δ 4.22 (m, 1H, C*H*OH); 3.72 (s, 3H, OC*H*₃); 2.56 (br s, 1H, O*H*); 2.46 (d, *J* 6.3 Hz, 2H, C*H*₂); 1.23 (d, *J* 6.1 Hz, 3H, C*H*₃CHOH).

Mass: calculated for $C_5H_{10}O_3$: *m/z* 118.06299, found [M]$^+$ 118.06286.

Conclusion

Hydrogenation with Noyori's catalyst required a customized rig for hydrogenation at high pressure to give good results for a large range of substrates. The Genêt modification does not require pressure and the catalyst can be prepared *in situ* which made the reaction very easy to carry out. Table 9.1 gives examples of the different substrates which can be hydrogenated by Noyori's[37] and Genêt's methods[35].

Table 9.1 Asymmetric hydrogenation of β-keto esters using [Ru(BiNAP)] complexes (results according to the relevant publications).

	[RuCl₂((R)-BINAP)][37] 70 to 100 atm. of H₂ Yield %, ee %	[RuBr₂((S)-BINAP)][35] 1 atm. of H₂ Yield %, ee %
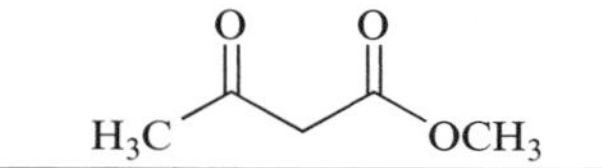 H_3C C(=O)CH₂C(=O) OCH_3	>99, >99	99, >98*
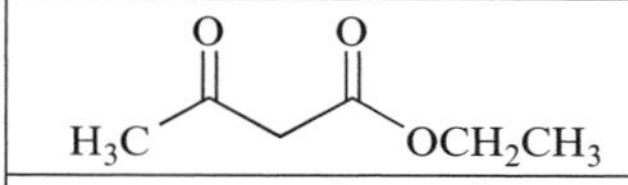 H_3C C(=O)CH₂C(=O) OCH_2CH_3	99, 99	–
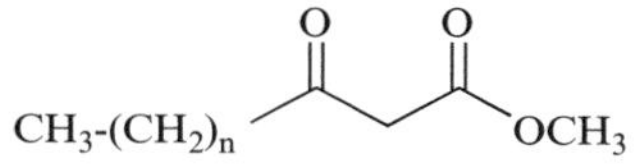 $CH_3\text{-}(CH_2)_n$ C(=O)CH₂C(=O) OCH_3	n=3; 99, 98	n=1; 100, 99 n=14; 100, 96
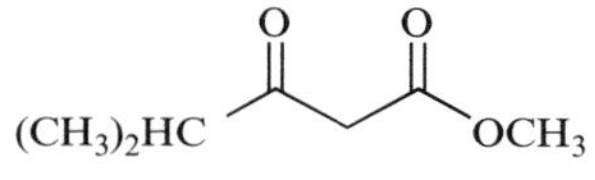 $(CH_3)_2HC$ C(=O)CH₂C(=O) OCH_3	99, >99	100, 97

* Rection validated

Alternatively, bis(phospholane) ligands can be very effective for the hydrogenation of carbonyl groups. Using a hydrogen pressure of only 60 psi, the hydrogenation of β-ketoesters by [Ru((R,R)–iPr-BPE)Br₂] (0.2 mol%) gives high catalytic efficiency (100 % conversion, 99 % enantiomeric excess) according to the literature[38]. This procedure (not validated in this volume) is similar to the one described later in this chapter for the hydrogenation of olefins using [(COD)Rh(Me-DuPHOS)] and [(COD)Rh(Me-BPE)] catalysts.

9.3 ASYMMETRIC TRANSFER HYDROGENATION OF β-KETOESTERS

KATHELYNE EVERAERE,[a] JEAN-FRANÇOIS CARPENTIER,[a] ANDRÉ MORTREUX[a] and MICHEL BULLIARD[b]

[a]Laboratoire de Catalyse, B.P. 108, 59652 Villeneuve d'Ascq Cedex, France.
[b]PPG-SIPSY, Z.I. La Croix Cadeau, B.P. 79, 49242 Avrillé Cedex, France.

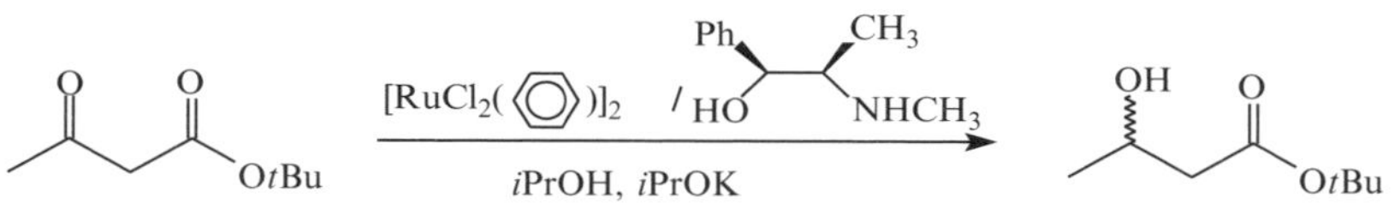

Materials and equipment

- Anhydrous 2-propanol, 19 mL
- *tert*-Butyl acetoacetate (98%), 316 mg, 2 mmol
- di-μ-Chlorobis[(benzene)chlororuthenium (II)], 5 mg, 0.01 mmol*
- (1S,2R)-(+)-Ephedrine, 6.6 mg 0.04 mmol
- Potassium 2-propylate solution, 0.12 mol.L^{-1}
- Ethyl acetate, distilled water
- Hydrochloric acid
- Sodium chloride
- Magnesium sulfate

- Two Schlenk tubes
- Magnetic stirrer plate
- Oil bath
- Vacuum line

Procedure

1. In a Schlenk tube equipped with a magnetic stirrer bar were placed under nitrogen [Ru(benzene)Cl$_2$]$_2$ (5 mg), ephedrine (6.6 mg) and dry 2-propanol (5 mL) previously degassed by three freeze–thaw cycles. The mixture was stirred for 20 minutes at 80 °C to give an orange solution which was allowed to cool to room temperature.

2. In a second Schlenk tube were placed under nitrogen *tert*-butyl acetoacetate and dry 2-propanol (14 mL). The mixture was degassed and added to the first solution. Finally, degassed potassium 2-propylate solution (1 mL) was added. The resulting orange solution was stirred at room temperature.

3. The reaction was monitored by GLC (BPX5 25 m × 0.32 mm column, 0.8 bar N$_2$, 60 °C).

4. After completion of the reaction (1 hour), the solution was neutralized with dilute hydrochloric acid and the solvent removed *in vacuo*. The residue was diluted with ethyl acetate and the organic solution was washed with saturated aqueous sodium chloride. The organic layer was dried over magnesium sulfate, concentrated under reduced pressure and distilled to afford *tert*-butyl 3-hydroxybutyrate (80%).

   The ee (44% in the S enantiomer) was determined by GLC (CHIRASIL-DEX CB 25 m × 0.25 mm column, 0.8 bar H$_2$, 85 °C); (S)-enantiomer: R_t 11.7 min, (R)-enantiomer: R_t 12.1 min.

   ^{1}H NMR (300 MHz, CDCl$_3$): δ (ppm) 4.02 (m, J 6 Hz, 1H, C$\underline{H}$(OH)); 3.5 (s, 1H, O$\underline{H}$), 2.2 (d, J 6 Hz, 2H, C$\underline{H}_2$); 1.35 (s, 9H, C(C$\underline{H}_3$)$_3$); 1.09 (d, J 6 Hz, 3H, C$\underline{H}_3$).

   ^{13}C NMR (300 MHz, CDCl$_3$): δ (ppm) 172.1 ($\underline{C}$OO); 80.9 ($\underline{C}$(CH$_3$)$_3$); 64.2 ($\underline{C}$H(OH)); 43.9 ($\underline{C}$H$_2$); 27.9 (C($\underline{C}$H$_3$)$_3$); 22.3 ($\underline{C}$H$_3$).[i]

* M. Bennett, A. Smith, *J. Chem. Soc., Dalton Trans.* 1974, 233.

Conclusion

This procedure offers a simple alternative for the reduction of β-ketoesters which does not require the use of an autoclave or hydrogen. The reaction is easily reproducible and leads to virtually quantitative yields of β-hydroxyesters under mild conditions. The use of sterically hindered esters, i.e. *iso*-propyl or *tert*-butyl β-ketoesters, significantly improves the catalytic activity, so that reactions go to completion in a reasonable time at room temperature (see below). When the reaction time is too long, transesterification may occur, giving rise to a mixture of an alkyl β-hydroxyester and the corresponding *iso*-propyl β-hydroxyester. Ephedrine as a chiral ligand affords modest to good enantiomeric excesses according to the nature of the β-ketoester but is essential for good activity of the catalytic system.

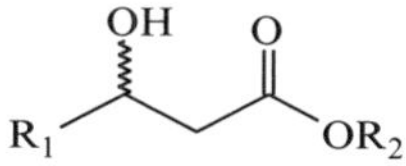

R_1	R_2	T(°C)	t (h)	GLC Yield (%)	ee (%)
Me	Et	20	10	100	39 (*S*)
Me	*i*Pr	20	4	100	40 (*S*)
Me	*t*Bu	20	1	100	44 (*S*)
Ph	Et	50	2.5	99	40 (*S*)
Ph	Et	50	15	85	94 (*S*)*

*Catalyst Precursor = [Ru(*p*-cymene)Cl$_2$]$_2$

9.4 (S,S)-1,2-BIS(*TERT*-BUTYLMETHYLPHOSPHINO)ETHANE (BisP*)[39]: SYNTHESIS AND USE AS A LIGAND

T. IMAMOTO

Department of Chemistry, Faculty of Science, Chiba University, Yayoi-cho, Inage-ku Chiba 263–8522, Japan, Phone & Fax: 81–43–290–2791, e-mail: imamoto@scichem.s. chiba-u.ac.jp

9.4.1 SYNTHESIS OF BisP*

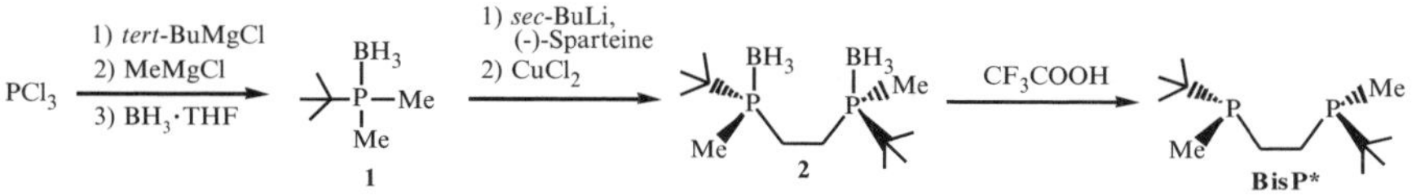

Materials and equipment

- Phosphorus trichloride, 6.4 g
- Dry tetrahydrofuran, 20 mL

- *tert*-Butylmagnesium chloride, 1.0 M THF solution, 52 mL
- Methylmagnesium chloride, 1.0 M THF solution, 112 mL
- Borane–THF complex, 1.0 M, 70 mL
- Hydrochloric acid
- Sodium chloride, sodium sulfate
- Silica gel
- *n*-Hexane, ethyl acetate, diethyl ether, toluene, ethyl alcohol
- (−)-Sparteine, 8.7 g
- *sec*-BuLi 1 M in cyclohexane, 37.1 mL
- Copper(II) chloride, 6.2 g
- Aqueous ammonia
- Trifluoromethanesulfonic acid
- Potassium hydroxide
- Basic alumina
- Argon

- Magnetic stirrer
- 300 mL Round-bottomed flask with a magnetic stirrer bar
- Dropping funnel
- Separatory funnel
- Dry ice
- Schlenk tube
- Oil bath
- Rotary evaporator
- Cannula

Procedure

1. To a solution of phosphorus trichloride (6.4 g) in dry tetrahydrofuran (20 mL) was added *tert*-butylmagnesium chloride (1.0 M THF solution, 52 mL) at −78°C under an argon atmosphere over a period of 2 hours.
2. After stirring at room temperature for 1 hour, methylmagnesium chloride (1.0 M THF solution, 112 mL) was added at 0 °C over 30 minutes. The reaction mixture was stirred at room temperature for 1 hour. To this solution, borane–THF complex (1.0 M, 70 mL) was added at 0 °C, and the mixture was stirred at the same temperature for 1 hour.
3. The reaction mixture was poured into cold 5% hydrochloric acid (200 mL). The organic layer was separated and aqueous layer was extracted with ethyl acetate (3 × 80 mL). The combined extracts were washed with brine, dried over sodium sulfate, and concentrated in vacuo.
4. The residual pasty solid was purified by column chromatography over silica gel (*n*-hexane : ethyl acetate = 8:1) to afford *tert*-butyl (dimethyl)phosphine–borane (1) 5.29 g (86%) as white crystals.
5. To a solution of (−)-sparteine (8.7 g) in diethyl ether (dry, 100 mL) *sec*-butyllithium (1 M in cyclohexane, 37.1 mL) was added at −78 °C under an

argon atmosphere. After stirring for 10 minutes, *tert*-butyl(dimethyl)phosphine–borane (1) (4.9 g) in diethyl ether (40 mL) was added, and the mixture was stirred at −65 °C for 3 hours. Dry copper (II) chloride (6.2 g) was added in one portion, and the mixture was gradually warmed to room temperature over 2 hours.

6. After stirring for 1 hour, 25 % aqueous ammonia (50 mL) was added dropwise. The organic layer was separated and aqueous layer was extracted with ethyl acetate (200 mL, twice). The combined extracts were washed with 5 % aqueous ammonia (100 mL), 2 M hydrochloric acid, and brine, and then dried over sodium sulfate, and concentrated *in vacuo*.

7. The residual solid was recrystallized from toluene twice to afford (*S,S*)-1,2-bis(boranato(*tert*-butyl)methylphosphino)ethane (2) 1.89 g (39 %) as white crystals $[\alpha]^{27}_D = -9.1°(c\ 1.0,\ CHCl_3)$.

8. In a Schlenk tube equipped with a magnetic stirrer bar were placed (*S,S*)-1,2-bis(boranato(*tert*-butyl)methylphosphino)ethane (2) (131 mg) and dry toluene (4 mL) under an argon atmosphere at 0 °C. To this solution, trifluoromethanesulfonic acid (0.22 mL) was added over a period of 5 minutes.

9. After stirring at 0 °C for 30 minutes and at room temperature for 1 hour, the solvent was removed *in vacuo*. The resulting pasty oil was dissolved in degassed ethanol/water (10/1) containing potassium hydroxide (280 mg), and stirred at 50 °C for 90 minutes. After cooling to room temperature, degassed diethyl ether (5 mL) was added, and the upper layer was collected through a cannula. The extraction was repeated three times. The combined extracts were dried over sodium sulfate. The solution was passed through a column of basic alumina (13 g) using degassed diethyl ether (30 mL) under an argon atmosphere.

10. The resulting solution was concentrated in vacuo to afford (*S,S*)-1,2-bis(*tert*-butylmethylphosphino)ethane (BisP*) as a solid or an oil.

9.4.2   SYNTHESIS OF 1,2-BIS(*TERT*-BUTYLMETHYLPHOSPHINO)ETHANERUTHENIUM BROMIDE[40](BisP*−Ru)

Materials and equipment

- (*S,S*)-1,2-Bis(*tert*-butylmethylphosphino)ethane (BisP*), 77 mg
- Bis(2-methylallyl)cyclooctadieneruthenium, 105 mg
- Argon

- *n*-Hexane (degassed) 2.5 ml
- Acetone (degassed) 11 ml
- 1.45 M HBr/Methanol

- Schlenk tube
- Magnetic stirrer
- Magnetic stirrer bar
- Oil bath
- Cannula
- Rotary evaporator

Procedure

1. In a Schlenk tube, equipped with a magnetic stirrer bar, were placed BisP* (77 mg), bis(2-methylallyl)cyclooctadieneruthenium (II) (105 mg), and degassed *n*-hexane (2.5 mL) under an argon atmosphere. The mixture was stirred at 60 °C for 10 hours.
2. To the reaction mixture, degassed *n*-hexane 5 mL was added. The solution was passed through a cannula capped with a filter paper, and concentrated *in vacuo*.
3. The resulting solid was dissolved in degassed acetone (11 mL). To this solution, 1.45 M methanolic hydrogen bromide (0.384 mL) was added dropwise. The mixture was stirred for two hours.
4. The mixture was filtered by passing through a cannula capped with a filter paper, and the solvent was removed *in vacuo* to afford 1,2-bis(*tert*-butylmethylphosphino)ethaneruthenium(II) bromide.

9.4.3 SYNTHESIS OF (*R*)-(−)-METHYL 3-HYDROXYPENTANOATE[40] USING (BisP*−Ru)

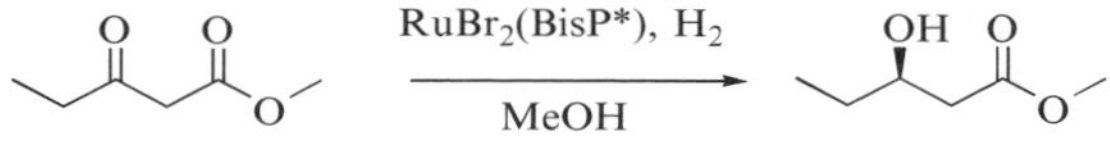

Materials and equipment

- Methanol–distilled water (10:1), 200 mL
- BisP*–RuBr₂ 52 mg,
- Methyl 3-oxopentanoate 10.7 g
- Argon, hydrogen
- Silica gel
- Chloroform, acetone

- Low pressure hydrogenator equipped with a 500 mL glass autoclave, a heater, and a magnetic stirrer

- Magnetic stirrer bar
- Rotary evaporator
- Chromatography column (200 mL)

Procedure

1. In a glass autoclave equipped with a magnetic stirrer bar were placed BisP*–RuBr$_2$ (52 mg), methyl 3-oxopentanoate (10.7 g), and degassed methanol/water (10/1) 200 mL under an argon atmosphere.
2. Then argon gas was replaced with hydrogen. The hydrogenation was performed at 70 °C under 6 kg/cm^2 of hydrogen for 10 hours.
3. The solvent was removed under reduced pressure and the residue was purified by chromatography on silica gel (chloroform:acetone = 3:1) to afford (R)-(−)-methyl 3-hydroxypentanoate (10.4 g, 96%, 98% ee) as a colourless oil.

    The ee (98%) was determined by HPLC (CHIRALCEL OD, flow rate 0.5 mL/min, eluent n-hexane/2-propanol = 95/5); (R)-enantiomer: R_t 13.8 min, (S)-enantiomer: R_t 28.4 min.

    ^{1}H NMR (400 MHz, CDCl$_3$): δ 0.97 (3H, t, J 7.4 Hz), 1.48–1.58 (2H, m), 2.41 (1H, dd, J 9.1 and 16.4 Hz), 2.53 (1H, dd, J 3.0 and 16.4 Hz), 2.91 (1H, br s), 3.72 (3H, s), 3.95 (1H, m).

    IR (neat, cm^{-1}): 3443 (OH), 2965, 2881, 1738 (C=O), 1439, 1284, 1172, 1113, 1068, 1015, 983.

Conclusion

The asymmetric hydrogenation with BisP*–RuBr$_2$ may be applied to a wide range of β-ketoesters, β-ketoamides, and β-ketophosphonates. Table 9.2 shows typical examples.

9.5 (1S, 3R, 4R)-2-AZANORBORNYLMETHANOL, AN EFFICIENT LIGAND FOR RUTHENIUM CATALYSED ASYMMETRIC TRANSFER HYDROGENATION OF AROMATIC KETONES

DIEGO A. ALONSO and PHER G. ANDERSSON

Department of Organic Chemistry, Uppsala University, Box 531, S-751 21 Uppsala, Sweden.

Asymmetric ruthenium-catalysed hydrogen transfer from 2-propanol to ketones is an efficient method for the preparation of optically active secondary alcohols[41]. Very recently, a new catalytic system has been developed based on ruthenium complexes having 2-azanorbornylmethanol as the chiral ligand (Figure 9.2), and their efficiency as catalysts for the enantioselective transfer hydrogenation of aromatic ketones has been demonstrated, affording the corresponding secondary alcohols with high rates and excellent ee's[42].

Table 9.2 Asymmetric hydrogenation of keto esters with BisP*–Ru(II) catalysts.

Entry	Substrate	BisP*	Product		Yield(%)	% ee
1	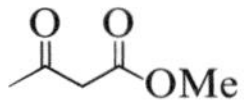 (methyl acetoacetate)	a	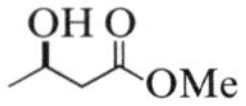 (methyl 3-hydroxybutanoate)		86	97
2[b]	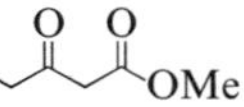 (methyl 3-oxopentanoate)	a	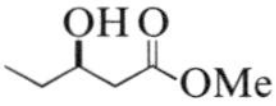 (methyl 3-hydroxypentanoate)		96	98
3		b			18	87
4	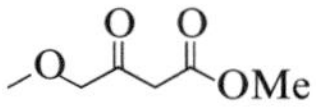 (methyl 4-oxo ester)	a	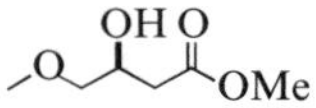		100	81
5	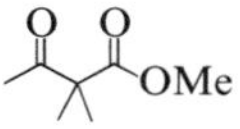 (methyl 2,2-dimethyl-3-oxobutanoate)	a	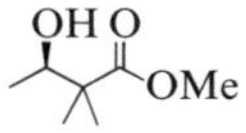		81	98
6[c]	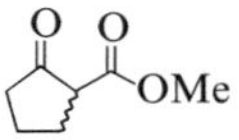 (methyl 2-oxocyclopentanecarboxylate)	a	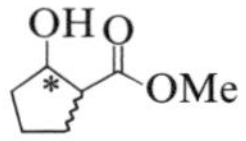 (*)	(syn)	13	91
				(anti)	70	96
7	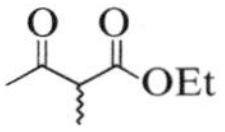 (ethyl 2-methyl-3-oxobutanoate)	a	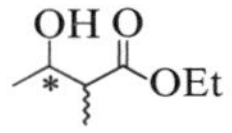 (*)	(syn)	39	96
				(anti)	45	97
8	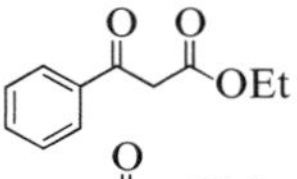 (ethyl benzoylacetate)	a	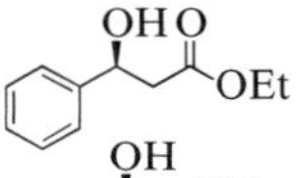		100	89
9	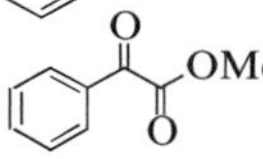 (methyl benzoylformate)	a	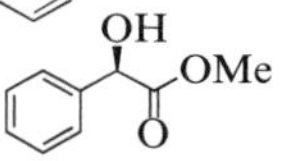		90	70
10	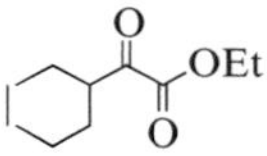 (ethyl cyclohexyl oxoacetate)	a	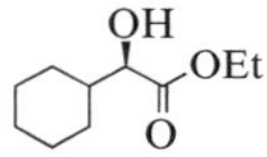		77	50
11	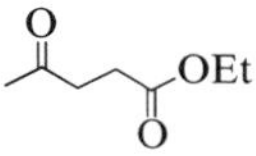 (ethyl 4-oxopentanoate)	a	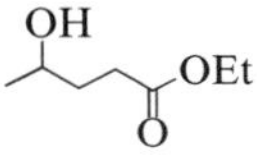		No reaction	
12	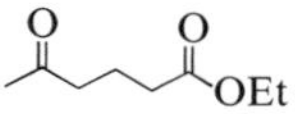 (ethyl 5-oxohexanoate)	a	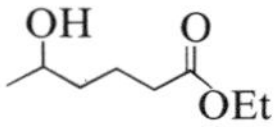		No reaction	
13[d]	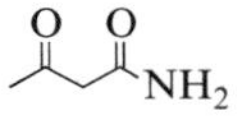 (3-oxobutanamide)	a	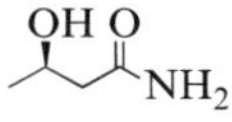		100	89
14[e]	 (dimethyl 2-oxopropylphosphonate)	a			51	85

a) Reaction conditions: substrate 2 mmol, catalyst/substrate = 0.005, 70 °C, 6 kg/cm^2.
b) Substrate 82 mmol, catalyst/substrate = 0.0013.
c) At 50 °C.
d) Substrate 0.7 mmol, catalyst/substrate = 0.014, at 50 °C.
e) At 60 °C.

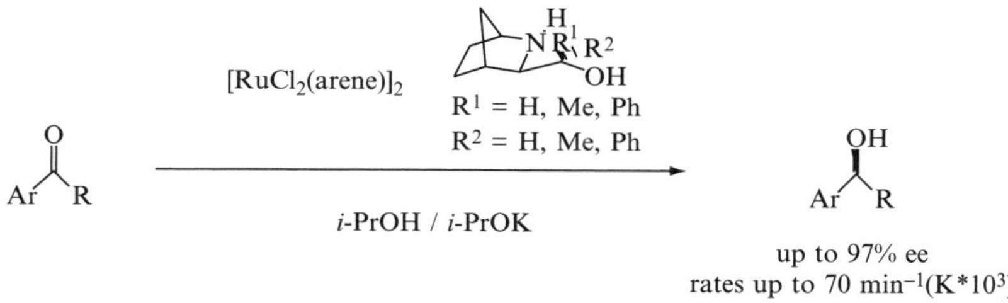

Figure 9.2 Transfer hydrogenation of ketones catalyzed by Ru(II)(2-azanorbornyl-methanol) complexes.

9.5.1 SYNTHESIS OF ETHYL (1S,3R,4R)-2-[(S)-1-PHENYLETHYLAMINO]-2-AZABICYCLO [2.2.1] HEPT-5-ENE-3-CARBOXYLATE

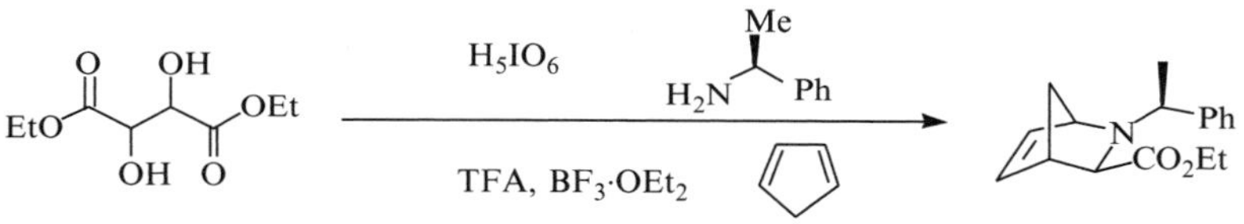

Materials and equipment

- Diethyl L-tartrate (98%), 10 g, 47.5 mmol
- *Ortho*-periodic acid (>99%), 10.8 g, 47.5 mmol
- (S)-(−)-α-Methylbenzylamine, (98%, 99% ee/GLC), 12.4 mL, 95 mmol
- Trifluoroacetic acid (99%), 7.4 mL, 95 mmol
- Boron trifluoride etherate (99%), 12.1 mL, 95 mmol
- Cyclopentadiene, 6.2 g, 95 mmol
- Triethylamine, 45 mL
- Dry diethyl ether, 30 mL
- Dry methylene chloride, 150 mL
- Pentane, diethyl ether
- Methylene chloride, 400 mL
- Saturated solution of sodium bicarbonate
- Anhydrous magnesium sulfate
- Activated molecular sieves, 3 Å, 0.4–0.8 mm beads, 17 g
- Celite®, 20 g
- Silica gel (Matrex 60 Å, 37–70 μm), 250 g
- TLC plates, SIL G-60 UV$_{254}$

- 100 mL and 250 mL round-bottomed flasks with magnetic stirrer bars
- Magnetic stirrer plate
- One glass sintered funnel, diameter 7 cm

- One 500 mL Erlenmeyer flask
- One Büchner funnel, diameter 10 cm
- One Büchner flask, 500 mL
- Filter paper
- One 500 mL separatory funnel
- One glass column, diameter 7 cm
- One Dewar flask
- Rotary evaporator

Procedure

1. Diethyl tartrate (10 g, 47.5 mmol) was placed in a 100 mL round-bottomed flask equipped with a magnetic stirrer bar, under nitrogen. Dry diethyl ether (30 mL) was then added and the mixture was cooled at 0 °C. To this solution, *ortho*-periodic acid (10.8 g, 47.5 mmol) was added carefully, portionwise, over 20 minutes and the resulting mixture was vigorously stirred for 1 hour under nitrogen at the same temperature.
2. Activated molecular sieves (7 g) were added to the reaction and stirring continued for 20 minutes.
3. The mixture was carefully filtered into a 250 mL round-bottomed flask through a thin pad of Celite. The filtration was completed by rinsing the packing with diethyl ether (50 mL).
4. The solvent was removed from the filtrate using a rotary evaporator to afford the corresponding ethyl glyoxylate (9.7 g, 95 mmol)
5. In the 250 mL round-bottomed flask equipped with a magnetic stirrer bar, the resulting ethyl glyoxylate (9.7 g) was placed under nitrogen. Dry methylene chloride (125 mL) was added followed by activated molecular sieves (10 g). The mixture was then cooled at 0 °C under nitrogen.
6. To this solution (S)-(−)-phenylethylamine (12.4 mL, 95 mmol) was added dropwise and when the addition was complete, stirring was continued for 1 hour at 0 °C.
7. The mixture was cooled to −78°C in a dry ice–acetone bath and trifluoroacetic acid (7.4 mL, 95 mmol), boron trifluoride etherate (12.1 mL, 95 mmol) and freshly distilled cyclopentadiene (6.2 g, 95 mmol) were added in that order over a 20 minute period. The mixture was stirred at −78°C for 5 hours before it was allowed to warm to room temperature.
8. The reaction was hydrolysed and neutralized (pH = 8) in a 500 mL Erlenmeyer flask with a saturated aqueous solution of sodium bicarbonate.
9. The resulting mixture was filtered into a Büchner funnel with the aid of a water aspirator and transferred to a 500 mL separatory funnel where the phases were separated. The aqueous layer was extracted with methylene chloride (2 × 150 mL). The combined organic layers were dried over magnesium sulfate, filtered and concentrated using a rotary evaporator to give the corresponding mixture of crude Diels–Alder adducts (minor *exo*-isomer + major *endo*-isomer + major *exo*-isomer. *Exo/endo* ratio: 98/2).

10. The major *exo* diastereoisomer was easily isolated from the crude reaction mixture by column chromatography (deactivated silica gel[43], pentane/ Et_2O: 99/1 to 80/20) to give, 14.1 g, 55% yield.

^1H NMR (200 MHz, $CDCl_3$): δ 7.34–7.19 (5H, m), 6.36–6.33 (1H, m), 6.22–6.18 (1H, m), 4.37 (1H, s), 3.87 (2H, q, *J* 6.8 Hz), 3.10 (1H, q, *J* 6.5 Hz), 2.96 (1H, s), 2.27 (1H, s), 2.20 (1H, d, *J* 8.4 Hz), 1.48 (3H, d, *J* 6.5 Hz), 1.01 (3H, d, *J* 7.0 Hz), 1.00 (1H, d, *J* 8.4 Hz).

This procedure has been scaled up to provide 28 g of the major Diels–Alder adduct.

9.5.2 SYNTHESIS OF (1*S*,3*R*,4*R*)-3-HYDROXYMETHYL-2-AZABICYCLO[2.2.1]HEPTANE

Materials and equipment

- Ethyl (1*S*,3*R*,4*R*)-2-[(*S*)-1-phenylethylamino]-2-azabicyclo[2.2.1]hept-5-ene-3-carboxylate, 10 g, 36.9 mmol
- 5% Pd–C, 2 g, 20 wt%
- Absolute ethanol (99%), 60 mL
- 95% Ethanol, 75 mL
- Hydrogen pressure (150 psi)
- Benzoyl chloride
- Triethylamine
- Lithium aluminium hydride, (97%), 8.4 g, 215 mmol
- Dry tetrahydrofuran, 130 mL
- 15 wt% NaOH aqueous solution, 8.4 mL
- Celite®, 20 g
- TLC plates, SIL G-60 UV_{254}

- One Büchner flask, 250 mL
- One glass sintered funnel, diameter 7 cm
- Two 250 mL round bottomed flasks, one equipped with a magnetic stirrer bar
- Magnetic stirrer
- Hydrogen pressure reactor vessel, 250 mL with stirrer
- Water aspirator
- Kugelrohr distillation equipment

Procedure

1. A solution of ethyl $(1S,3R,4R)$-2-[(S)-1-phenylethylamino]-2-azabicyclo [2.2.1]hept-5-ene-3-carboxylate (major *exo*-Diels–Alder adduct) (10 g, 36.9 mmol) in absolute ethanol (60 mL) was stirred under a hydrogen pressure of 150 psi at room temperature for 48 hours in the presence of activated 5 % Pd–C (2 g, 20 wt%).

2. The Pd–C catalyst was then removed by filtration through Celite in a sintered glass funnel with the aid of a water aspirator. The filtration was completed by rinsing the packing with 95 % ethanol (75 mL). **Attention: due to the pyrophoric properties of hydrogen-saturated palladium, it is important to keep the filter plug under a layer of ethanol.**

3. The filtrate was transfered to a 250 mL round bottomed flask and the solvent was removed using a rotary evaporator to afford the corresponding pure amino ester as a pale yellowish oil (6.11 g, 98 % yield).

   The ee (98 %) of the ligand was determined at this point as its N-benzoyl derivative [44] by HPLC (ChiralCelOD-H, 254 nm UV detector, flow 0.4 mL/ min, eluent hexane/i-PrOH: 80/20); R_t (minor) 15.0 min, R_t (major) 22.9 min.

   ^{1}H NMR (400 MHz, CDCl$_3$): δ 1.21 (1 H, br d, J 9.8 Hz), 1.25 (3 H, t, J 7.1 Hz), 1.34–1.66 (5 H, m), 2.18 (1 H, br s), 2.59, 3.27, 3.50 (1 H each, 3 br s) and 4.15 (2 H, q, J 7.1 Hz).

4. Lithium aluminium hydride (8.4 g, 215 mmol) was placed in a 250 mL round bottomed flask equipped with a magnetic stirrer bar and flushed with nitrogen. Dry tetrahydrofuran (120 mL) was added and the suspension was cooled with the aid of an ice-bath.

5. A solution of the amino ester (6.11 g, 36.2 mmol) in dry tetrahydrofuran (10 mL) was carefully, dropwise added to this supension. When the addition was complete, the mixture was stirred for 30 minutes at room temperature.

6. After completion according to TLC, the reaction was quenched adding consecutively 8.4 mL of water, 8.4 mL of 15 % NaOH solution and 25.2 mL of water. The mixture was then stirred for 30 min at room temperature.

7. The aluminates were removed by filtration through a sintered glass funnel with the aid of the water aspirator. The filtration was completed by rinsing the packing with tetrahydrofuran (75 mL).

8. Solvent evaporation in a rotary evaporator afforded the amino alcohol product (4.14 g, 90 % yield). The purity of the crude product is high enough to be used in the catalysis experiments, but it can be purified further by vacuum distillation in a Kugelrohr [90–100 °C, 0.030–0.035 mbar] cooling the recipient flask with dry ice (84 % yield from the amino ester, as white needles).

   ^{1}H NMR (400 MHz, CDCl$_3$): δ 1.15 (1 H, dt, J 9.8, 1.4 Hz), 1.40–1.30 (2 H, m), 1.72–1.53 (3 H, m), 2.17 (1 H, m), 2.89–2.79 (3 H, m), 3.17 (1 H, dd, J 10.7, 8.2 Hz), 3.40 (1 H, dd, J 10.7, 5.4 Hz) and 3.43 (1 H, br s).

9.5.3 RUTHENIUM-CATALYZED ASYMMETRIC TRANSFER HYDROGENTION OF ACETOPHENONE

Materials and equipment

- Anhydrous isopropanol, 20 mL
- Acetophenone (99 %), 235 μl, 2 mmol
- [RuCl$_2$(p-cymene)]$_2$, 3.06 mg, 0.005 mmol (0.25 mol%)
- (1S,3R,4R)-3-Hydroxymethyl-2-azabicyclo[2.2.1]heptane, 5.08 mg, 0.04 mmol (2 mol%)
- 1 M Solution of i-PrOK in i-PrOH, 50 μl, 0.05 mmol
- 1 M solution of HCl, two drops
- Ethyl acetate, pentane
- Celite®, 5 g
- Silica gel (Matrex 60 Å, 37–70 μm), 13 g

- One 10 mL two-necked round-bottomed flask equipped with magnetic stirrer bar
- Magnetic stirrer with thermostatically controlled oil bath and thermometer
- Reflux condenser
- One 30 mL Schlenk flask equipped with magnetic stirrer bar
- Magnetic stirrer
- Glass sintered funnel, diameter 3.5 cm
- One 50 mL filter flask
- Rotary evaporator
- Chromatography column, diameter 2.5 cm

Procedure

1. The 10 mL two-necked round-bottomed flask, equipped with a magnetic stirrer bar, was dried in a oven at 120 °C for 10 hours. The flask was removed, sealed, cooled under vacuum and flushed with nitrogen.
2. The ruthenium complex dimer (3.06 mg, 0.25 mol%) and the chiral ligand (5.08 mg, 2 mol%) were then weighed into the round-bottomed flask and any moisture was azeotropically removed *via* evaporation of benzene (5 × 5 mL) at reduced pressure.
3. A condenser was attached to the flask which was sealed under vacuum and flushed with nitrogen. The residue was dissolved in dry (freshly distilled from CaCl$_2$) i-PrOH (5 mL). The solution was refluxed under nitrogen for 30 minutes before it was cooled to room temperature.

4. Once the clear reddish solution of the catalyst had cooled to room temperature, it was transferred under a gentle stream of nitrogen to a Schlenk flask containing a solution of the ketone (235 μL, 2 mmol) and potassium isopropoxide (0.05 mmol, 2.5 mol%) in i-PrOH (15 mL). The resulting solution was then stirred for 1.5 hours at room temperature under nitrogen (monitored by GC and/or ^{1}H NMR).

5. After completion, the reaction was neutralized with a 1 M solution of HCl (2 drops) and concentrated *in vacuo* to give the crude product. After dilution with ethyl acetate and removal of the catalyst by filtration over a thin pad of Celite, the sample was analysed.

　　　The ee of the alcohol (94 %) was determined by HPLC analysis (ChiralCel OD-H) using a 254 nm UV detector and a flow rate of 0.5 mL/min of 5 % of i-PrOH in hexane. (S)-α-Methylbenzyl alcohol (major): R_t 20.65 min; (R)-α-methylbenzyl alcohol (minor): R_t 17.72 min.

6. Finally, the crude product was purified by flash chromatography (eluent: pentane/ethyl acetate: 85/15) to afford 224.5 mg of pure alcohol (92 % yield).

Conclusion

The procedure is very easy to reproduce and the asymmetric transfer hydrogenation may be applied to a wide range of aromatic ketones. Table 9.3 gives different substrates that can be reduced with the Ru(II)-(2-azanorbornylmethanol) complex in *iso*-propanol

Table 9.3 Hydrogen transfer reduction of ketones using Ru (II)-(*1S, 3R, 4R*)-3-hydroxymethyl-2-azabicyclo [2.2.1]heptane as catalyst.

			Product	
Entry	Ar	R	Yield (%)	ee %
1	Naphthyl	Me	98	97 (S)
2	Ph	n-C$_4$H$_9$	80	95 (S)
3	m-Tol	Me	94	94 (S)
4	m-MeO-C$_6$H$_4$	Me	96	94 (S)
5	m-NH$_2$-C$_6$H$_4$	Me	100	93 (S)

REFERENCES

1. Noyori, R., Hashiguchi, S. *Acc. Chem. Res.*, 1997, **30**, 97.
2. Blaser, H.-U. *Tetrahedron: Asymmetry*, 1991, **2**, 843.
3. Izumi, Y. *Advance Catalysis*, 1983, **32**, 215.
4. Knowles, W.S., Sabacky, M.J. *J. Am. Chem. Soc., Chem. Commun.*, 1968, 1445.

5. Horner, L., Siegel, H., Büthe, H. *Angew. Chem. Int. Ed. English*, 1968, **7**, 942.
6. Singh, V.K. *Synthesis* 1992, 605.
7. Wallbaum, S., Martens, J. *Tetrahedron: Asymmetry*, 1992, **3**, 1475.
8. Buono, G., Chiodi, O., Wills, M. *Synlett*, 1999, 377.
9. Gamble, M.P., Smith, A.R.C., Wills, M. *J. Org. Chem.*, 1998, **63**, 6068.
10. Bolm, C. *Angew. Chem. Int. Ed. English*, 1993, **32**, 232.
11. Fiaud, J.C., Kagan, H.B. *Bulletin de la Societe Chimique Francaise*, 1969, 2742.
12. Hirao, A., Itsuno, S., Nakahama, S., Yamasaki, N. *J. Am. Chem. Soc., Chem. Commun.*, 1981, 315.
13. Brunel, J.M., Pardigon, O., Faure, B., Buono, G. *J. Am. Chem. Soc., Chem. Commun.*, 1992, 287.
14. Bertschy, H., Chenault, H.K., Whitesides, G.M. *Enzyme Catalysis in Organic Synthesis*; VCH: Weinheim, New York, 1995; Vol. II.
15. Chan, A.S., Chen, C.-C, Lin, C.-W., Lin, Y.-C., Cheng, M.-C., Peng, S.-M. *J. Am. Chem. Soc., Chem. Commun.*, 1995, 1767.
16. Uematsu, N., Fujii, A., Hashiguchi, S., Ikariya, T., Noyori, R. *J. Am. Chem. Soc.*, 1996, **118**, 4916.
17. Oppolzer, W., Wills, M., Starkemann, C., Bernadinelli, G. *Tetrahedron Lett.*, 1990, **31**, 4117.
18. Tillack, A., Lefeber, C., Peulecke, N., Thomas, D., Rosenthal, U. *Tetrahedron Lett.*, 1997, **38**, 1533.
19. Willoughby, C.A., Buchwald, S.L. *J. Am. Chem. Soc.*, 1994, **116**, 8952.
20. Lee, N.E., Buchwald, S.L. *J. Am. Chem. Soc.*, 1994, **116**, 5985.
21. Sugi, K.D., Nagata, T., Yamada, T., Mukaiyama, T. *Chem. Lett.*, 1997, 493.
22. Cheong Chan, Y.N., Osborn, J.A. *J. Am. Chem. Soc.*, 1990, **112**, 9400.
23. Cheong Chan, Y.N., Meyer, D., Osborn, J.A. *J. Chem. Soc., Chem. Commun.*, 1990, 869.
24. Spindler, F., Pugin, B., Blaser, H.-U. *Angew. Chem. Int. Ed. English*, 1990, **102**, 558.
25. Schnider, P., Koch, G., Pretot, R., Wang, G., Bohnen, F.M., Kruger, C., Pfaltz, A. *Chem. Eur. J.* 1997, **3**, 887.
26. Buriak, J.M., Osborn, J.A. *Organometallics* 1996, **15**, 3161.
27. Bakos, J., Orosz, A., Heil, B., Laghmari, M., Lhoste, P., Sinou, D. *J. Am. Chem. Soc., Chem. Commun.*, 1991, 1684.
28. Itsuno, S., Sakurai, Y., Ito, K., Hirao, A., Nakayama, S. *Bull. Chem. Soc. Jpn.* 1987, **60**, 395.
29. Noyori, R., Takaya, H., Ohta, T. *Asymmetric Hydrogenation*; Ojima, I., Ed., VCH: Weinheim, 1993, pp 1.
30. Noyori, R. *Acta Chemica Scandinavia* 1996, **50**, 380.
31. Buono, G., Chiodi, O., Wills, M. *Synlett* 1999, 377.
32. Wallbaum, S., Martens, J. *Tetrahedron: Asymmetry* 1992, **3**, 1475.
33. Bertschy, H., Chenault, H.K., Whitesides, G.M. *Enzyme Catalysis in Organic Synthesis*; VCH: Weinheim, New York, 1995; Vol. II.
34. Genêt, J.P., Pinel, C., Ratovelomanana-Vidal, V., Mallart, S., Pfister, X., Bischoff, L., Caño de Andrade, M.C., Darses, S., Galopin, C., Laffitte, J.A. *Tetrahedron: Asymmetry* 1994, **5**, 675.
35. Genêt, J.P., Ratovelomanana-Vidal, V., Caño de Andrade, M.C., Pfister, X., Guerreiro, P., Lenoir, J.Y. *Tetrahedron Lett.*, 1995, **36**, 4801.
36. Leonard, J., Lygo, B., Procter, G. *Advanced Practical Organic Chemistry*; Second Edition ed., Blackie Academic and Professional:, 1995.

37. Noyori, R., Ohkuma, T., Kitamura, M., Takaya, H. *J. Am. Chem. Soc.*, 1987, **109**, 5856.
38. Burk, M.J., Harper, T.G.P., Kalberg, C.S. *J. Am. Chem. Soc.*, 1995, **117**, 4423.
39. Imamoto, T., Watanabe, J.,Wada, Y., Masuda, H., Yamada, H., Tsuruta, H., Matsukawa, S., and Yamaguchi, K. *J. Am. Chem. Soc.*, 1998, **120**, 1635–6.
40. Yamano, T., Taya, N., Kawada, M., Huang, T., Imamoto, T. *Tetrahedron Lett.*, 1999, **40**, 2577–80.
41. (a) Sasson, Y., Blum, J. *Tetrahedron Lett.* 1971, **24**, 2167. (b) Sasson, Y., Blum, J. *J. Org. Chem.* 1975, **40**, 1887. (c) Linn, D.E., Halpern, G. *J. Am. Chem. Soc.* 1987, **109**, 2969. (d) Noyori, R., Takaya, H. *Acc. Chem. Res.* 1990, **23**, 345. (e) Chowdhury, R.L., Bäckvall, J.E. *J. Chem. Soc. Chem., Commun.* 1991, 1063. (f) Gamez, P., Dunjic, B., Lemaire, M. *J. Org. Chem.* 1996, **61**, 5196. (g) Palmer, M., Walsgrove, T., Wills, M. *J. Org. Chem.* 1997, **62**, 5226. (h) Jiang, Y., Jiang, Q., Zhang, X. *J. Am. Chem. Soc.* 1998, **120**, 3817. (i) Nishibayashi, Y., Takei, I., Uemura, S., Hidai, M. *Organometallics* 1999, **18**, 2291. (j) For a recent review, see: Palmer, M.J., Wills, M. *Tetrahedron: Asymmetry* 1999, **10**, 2045.
42. (a) Alonso, D.A., Guijarro, D., Pinho, P., Temme, O., Andersson, P.G. *J. Org. Chem.* 1998, **63**, 2749. (b) Alonso, D.A., Nordin, S.J.M., Brandt, P., Andersson, P.G. *J. Am. Chem. Soc.* 1999, **121**, 9580.
43. When mentioned, deactivated silica gel means that it was treated with 5% triethylamine in pentane and the column was eluted with the same solvent mixture until the outflowing eluent was basic according to pH paper.
44. The NH ester was *N*-benzoylated in methylene chloride by reaction with benzoyl chloride in the presence of triethylamine.

10 Asymmetric Reduction of Ketones Using Bakers' Yeast

CONTENTS

10.1   BAKERS' YEAST REDUCTION OF ETHYL ACETOACETATE

Among the strategies developed for the preparation of optically active ester derivatives, the bakers' yeast reduction of the corresponding β-ketoesters is one of the most useful methods. Because of its low cost, ready availability and its utility, bakers' yeast can be considered as a relatively simple reagent which is very easy to handle[1–3].

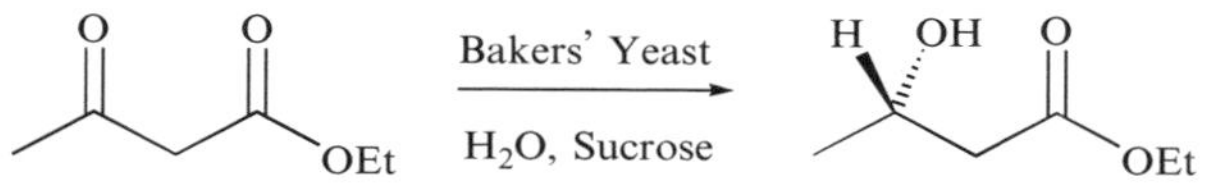

Materials and equipment

- Ethyl acetoacetate, 9.8 mL, 10.0 g, 77 mmol
- Bakers' yeast (*Saccharomyces cerevisiae*), Sigma Type II, 20 g
- Sucrose, commercially available table sugar, 150 g
- Celite® or hyflo super cell, 20 g
- Tap water, 650 mL
- Sodium chloride
- Ethyl acetate, dichloromethane
- *p*-Anisaldehyde

- Magnesium sulfate

- 2 L three-necked round-bottomed flask
- Bubbler
- Thermostatically controlled oil bath and thermometer
- Orbital shaker at 30 °C and 220 r.p.m. (optional)
- Wide sinter funnel
- Magnetic stirrer plate
- Rotary evaporator
- Kugelrohr apparatus

Procedure

1. A 2 L three-necked round-bottomed flask equipped with a bubbler and a thermometer was charged with tap water (400 mL), sucrose (75 g) and dried yeast (10 g), added in this order. The mixture was stirred very gently (150 r.p.m.).

   After 1 hour, carbon dioxide should be evolved at approximately 1–2 bubbles/second. Alternatively the whole reaction may be carried out in a 2 L conical-flask placed in an orbital shaker at 30 °C and 220 r.p.m.
2. Ethyl acetoacetate (5 g) was then added dropwise to the fermenting solution and the mixture stirred at ambient temperature for 24 hours.
3. A warm solution (40 °C) of sucrose (50 g) in tap water (250 mL) was then added and the mixture stirred for 1 hour before a further aliquot of ethyl acetoacetate (5 g) was added. The mixture was then stirred for a further 18 hours.
4. The reaction was followed by TLC (eluent: dichloromethane). The starting material was UV active, stained yellow with *p*-anisaldehyde, R_f 0.75.

   At this stage a very slight trace of ethyl acetoacetate could be seen by TLC and to assist further reduction more sucrose (25 g) and yeast (10 g) were added. The reaction was then stirred at 30 °C for 18 hours.
5. When no more starting material was apparent by TLC, the reaction was considered to be complete.

   It is essential that all the starting material is consumed before terminating the reaction.
6. Celite® or hyflo super cell (20 g) was added to the suspension which was then filtered through a pad of hyflo in a wide-sinter funnel; the pad was washed with water (100 mL). The filtrate was saturated with sodium chloride and then extracted with ethyl acetate (5 × 500 mL). The combined extracts were dried over magnesium sulfate, filtered and the solvent was removed under reduced pressure to afford a pale viscous oil.
7. The crude product was then distilled using Kugelrohr apparatus (56 °C, 12 mmHg) to afford the desired alcohol as a clear colourless oil (5.80 g, 57 %).
   $[\alpha]^{25}_D + 35.4°$ (c 1.35, CHCl$_3$), [Lit. $[\alpha]^{25}_D + 37.2°$ (c. 1.3, CHCl$_3$)] which corresponds to 80–85 % ee.

NMR ^{1}H (200 MHz, CDCl$_3$): δ 4.18 (m, 1H, CH); 4.17 (q, J 7.15 Hz, 2H, CH_2CH$_3$); 3.17 (brs, 1H, exchangeable with D$_2$O, OH); 2.45 (d, J 7 Hz, 2H, CH_2); 1.29 (t, J 7.15 Hz, 3H, CH$_2$CH_3); 1.23 (d, J 5.5 Hz, 2H, CHCH_3). IR (Thin film, cm^{-1}): 3440, 2980, 1730, 1300.

Conclusion

In the original paper, the authors performed the reaction using commercially available bakers' yeast from a supermarket or bakery. Initially a trial run using similar quantities of Sigma dried yeast resulted in an extremely vigorous initial fermentation, so the quantity of dry yeast was reduced by factor of 5. The contributors assessed the enantiomeric excess of the alcohol by formation of the (+)-MTPA ester and examination of the ^{19}F NMR spectrum. However, the value obtained for the optical rotation was consistent with that reported in the literature.

Bakers' yeast is an inexpensive and readily available reducing agent that can be considered as a relatively simple reagent which is very easy to handle. Different substrates which can be reduced by bakers' yeast are reported in Table 10.1[1]; some other methods are described in a previous publication[4].

Table 10.1 Reduction of carbonyl compounds mediated by bakers' yeast (results according to the literature).

	Yield %	ee %
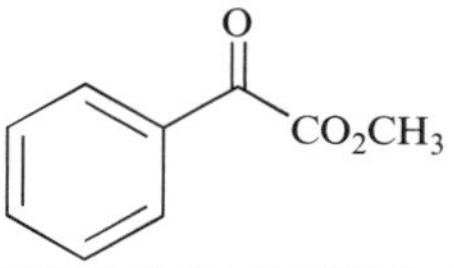 benzoylformate methyl ester (Ph–C(=O)–CO$_2$CH$_3$)	59	100
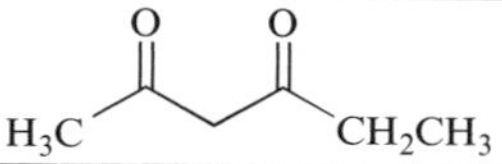 H$_3$C–C(=O)–CH$_2$–C(=O)–CH$_2$CH$_3$	48–100 57*	>97 Ca 85*
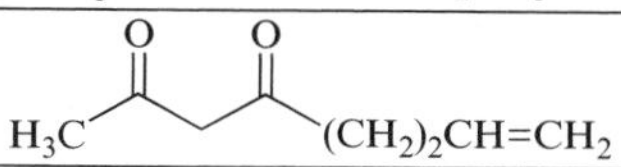 H$_3$C–C(=O)–CH$_2$–C(=O)–(CH$_2$)$_2$CH=CH$_2$	100	>99
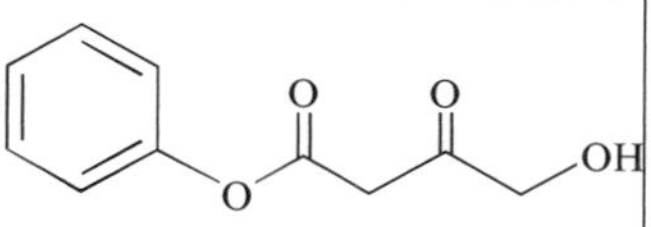 PhO–C(=O)–CH$_2$–C(=O)–CH$_2$OH	99	>99
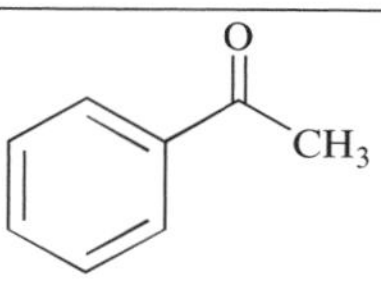 Ph–C(=O)–CH$_3$	45	89

* Reaction validated

10.2 ENANTIOSELECTIVE SYNTHESIS OF (Z)-N-CARBOBENZYLOXY-3-HYDROXYPROLINE ETHYL ESTER

MUKUND P. SIBI and JAMES W. CHRISTENSEN

Department of Chemistry, North Dakota State University, Fargo ND, 58105-5516, Sibi@-plains.nodak.edu

10.2.1 IMMOBILIZATION OF BAKERS' YEAST

Materials and equipment

- Supermarket variety bakers' yeast, 20 g[5]
- Sodium alginate, 5 g
- Water, 200 mL
- 10 % (w/v) Calcium chloride, 670 mL

- 2 L Erlenmeyer flask
- Magnetic stirrer
- 1 L Dropping funnel

Procedure

1. Two separate solutions of Red Star active dry yeast (or other supermarket variety, 20 g) and sodium alginate (5 g) each in water (200 mL) were prepared by very slow addition of the respective reagent to the rapidly stirred solvent.
2. When each solution became an homogeneous viscous fluid, they were combined and added dropwise via an addition funnel to calcium chloride (670 mL of 10 % (w/v)). The droplets formed gelatinous beads upon impact with the salt solution. The size and shape of the beads may be adjusted by the rate of addition.
3. The beads were washed five times with water (portions of 500 mL), and used immediately in the reduction of ketone.

10.2.2 BAKERS' YEAST REDUCTION OF (Z)-N-CARBOBENZYLOXY-3-KETOPROLINE ETHYL ESTER

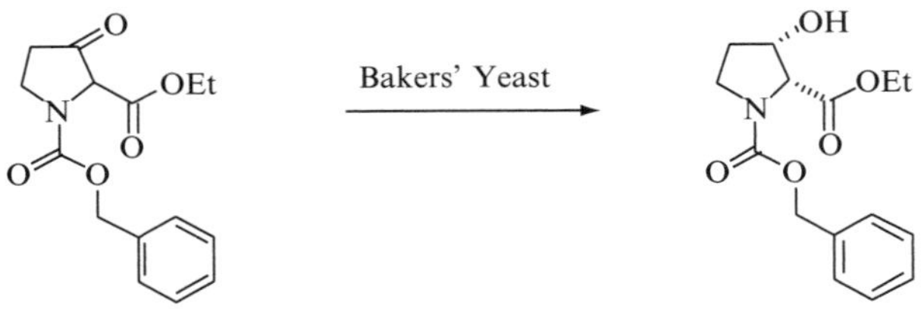

Materials and equipment

- Immobilized Bakers' Yeast, 200 g
- Sucrose, 40 g
- N-CBZ-3-Ketoproline ethyl ester, 5.64 g
- Ethanol, 10 mL
- Celite

- 2 L Erlenmeyer flask
- Magnetic stirrer
- Dropping funnel
- Buchner funnel

Procedure

1. The beads from the previous procedure ($\sim$200 g) were placed into a 2 L Erlenmeyer flask containing a large magnetic bar. Distilled water was added to give $\sim$950 mL total volume. Sucrose (40 g) was added and the mixture was allowed to stir vigorously for 30 minutes after which time a solution of the keto ester (5.64 g, 20 mmol) in ethanol (10 mL), was added over a period of $\sim$3 hours. The reaction was monitored by TLC and stopped when most of the starting material had been consumed (no longer than 24 hours).

2. The aqueous liquid was decanted onto a bed of Celite in an 11 cm Büchner funnel. The remaining yeast beads were washed and decanted three times using ethyl acetate (200 mL portions). The beads were then pressed free of their liquid content by compression against the filter. The resulting yeast cake was rinsed with ethyl acetate ($\sim$100 mL) and the layers of the filtrate were then separated. The aqueous layer was saturated with sodium chloride, filtered through Celite to remove gelatinous emulsions, and then extracted with ethyl acetate (200 mL). The combined organics were dried over magnesium sulfate, filtered, and evaporated to give 5.53 g of the crude product.

3. The crude product may be purified by chromatography over silica gel (eluted with 1:1 ethyl acetate–hexane). Average purified yield $\sim$85%. R_f 0.35 (50:50 hexane:ethyl acetate).

   The optical purity and stereochemistry was established by conversion to the known N-BOC protected proline and by comparison of spectral data and rotation values[6]. Additionally, the optical purity was established by HPLC analysis of the 3,5-dinitrobenzoates on a Pirkle column and by Mosher ester analysis.

   ^{1}H NMR (400 MHz, CDCl$_3$): δ 1.24–1.30 (m, 3H), 2.02–2.12 (m, 2H), 2.65 (s, 1H, broad), 3.53–3.58 (m, 1H), 3.69–3.73 (m, 1H), 4.08–4.15 (m, 1H), 4.22–4.26 (m, 1H), 4.41–4.46 (m, 1H), 4.59–4.62 (m, 1H), 5.04–5.19 (m, 2H), 7.29–7.36 (m, 5H).

   ^{13}C NMR (100 MHz, CDCl$_3$): δ 14.0 and 14.1, 32.0 and 32.8, 44.2 and 44.4, 61.2 and 61.3, 63.5 and 63.8, 67.1, 71.4 and 72.2, 127.7 and 127.8, 127.9

and 128.0, 128.3 and 128.4, and 136.3 and 136.5, 154.4 and 154.8, 169.8 and 170.0 (resonances are doubled due to hindered rotation).

Rotation was recorded on a JASCO-DIP-370 instrument: $[\alpha]_D^{26} + 21.87$ (c 1.996, CH_2Cl_2).

Conclusions

The reduction of β-keto esters by bakers' yeast is a convenient method for establishing multiple chiral centres in a single step. The procedure is very simple and can be carried out on a large scale. Isolation of the product is often problematic if one uses the yeast directly. The procedure using immobilized yeast allows for easy work up and higher chemical yield[7]. The present work is a modification of the procedure for the reduction of N-BOC-3-ketoproline ethyl ester originally reported by Knight et al.[8] The product hydroxyproline has served as a starting material in the synthesis of the indolizidine alkaloid slafra-mine[9].

REFERENCES

1. Bertschy, H., Chenault, H.K., Whitesides, G.M. *Enzyme Catalysis in Organic Synthesis*; VCH: Weinheim, New York, 1995; Vol. II.
2. Servi, S. *Synthesis* 1990, 1.
3. Csuk, R., Glänzer, B.I. *Chemical Reviews* 1991, **91**, 49.
4. Roberts, S.M. *Preparative Biotransformations* Wiley, Chichester, 1997.
5. Any supermarket variety is applicable.
6. Cooper, J., Gallagher, P.T., Knight, D.W. *J. Chem. Soc., Chem. Commun.*, 1988, 509.
7. Nakamura, K., Kawai, Y., Oka, S., Ohno, A. *Tetrahedron Lett.*, 1989, **30**, 2245 and references therein.
8. This procedure utilizes N-BOC-3-ketoproline ethyl ester as the substrate: Knight, D.W., Sibley, A.W. *J. Chem. Soc., Perkin Trans.* 1, 1997, 2179.
9. (a) Sibi, M.P., Christensen, J.W., Li, B., Renhowe, P. *J. Org. Chem.*, 1992, **57**, 4329.
   (b) Sibi, M.P., Christensen, J.W. *J. Org. Chem.*, 1999, **64**, 6434.

11 Asymmetric Reduction of Ketones Using Nonmetallic Catalysts

CONTENTS

11.1 INTRODUCTION

The amino alcohol–borane complex used in asymmetric reduction often consists of a boron hydride with one of a variety of chiral ligands based on vicinal amino alcohols derived from the corresponding amino acids (Figure 11.1). The complex is made by ligand exchange on treating a solution of amino alcohol with borane–tetrahydrofuran (BH_3.THF) or borane–dimethylsulfide (BH_3. SMe_2) complexes. The basicity of the nitrogen of the oxazaborolidine is considerably reduced, the boron is only loosely bound to the nitrogen.

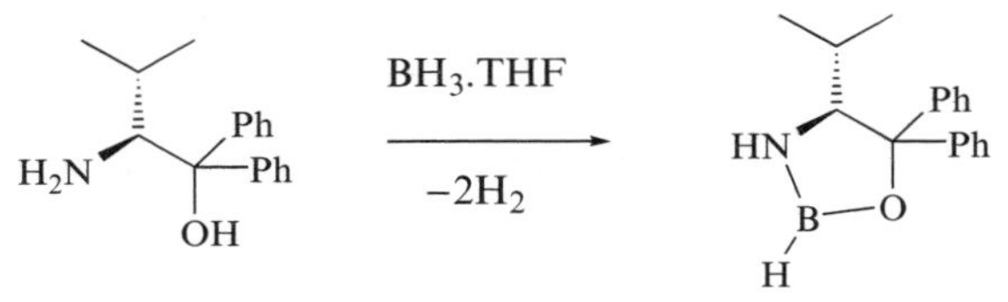

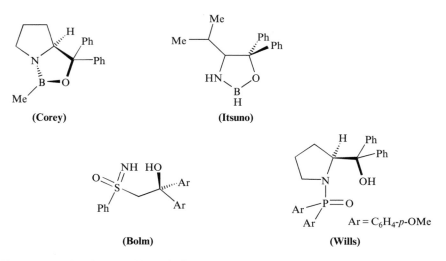

Figure 11.1 Complexation of borane with valinol[1].

Other nonmetallic catalysts were found to reduce ketones with high enantiomeric excess such as oxazaphospholidines (phosphorus analogues of oxazaborolidines), which were synthesized from (S)-prolinol and phenyl bis(dimethylamino)phosphine[2,3] Oxazaphosphinamide complexes, derived from oxazaphospholidines, react with borane to give a heterocycle in which the borane is activated by a strong donation from the oxygen atom of the N–P=O system coupled with a weaker interaction of the substrate carbonyl lone pair with the phosphorus atom[4]. Hydroxysulfoximines react with borane to give a six-membered heterocycle. The phenyl group and the electronic properties of the sulfoximine oxygen direct the coordination of the ketone[5].

Some of the above-mentioned catalysts or precursors are commercially available, such as the Corey catalyst (S)-3,3-diphenyl-1-methyltetrahydro-3H-pyrrolo[1,2-c][1,3,2]oxazaborole (Me-CBS). The amino alcohol (S)-(−)-2-amino-3-methyl-1,1'-diphenylmethan-1-ol, used as the ligand in the Itsuno catalyst is also readily available. The ligand used to prepare the oxazaphospholidine or oxazaphosphinamide complex (from Wills) can be synthesized easily from commerically available material. The preparation of the Bolm β-hydroxysulfoximine catalyst will be described in this chapter (Figure 11.2).

Figure 11.2 Catalysts and ligands for carbonyl reduction by borane.

As an example of a typical catalytic cycle, Figure 11.3 shows a mechanism suggested by Corey[6]. The reduction occurs by co-ordination of the oxazaborolidine electrophilic boron and the carbonyl oxygen. Then hydrogen transfer occurs from the amino borohydride anion unit (NBH_3^-) to the activated carbonyl via a six-membered ring transition state. Subsequent ligand exchange to form the alkoxy borane followed by displacement completes the catalytic cycle. For oxazaphosphinamide and hydroxysulfoximine catalysts, similar catalytic cycles have been suggested.

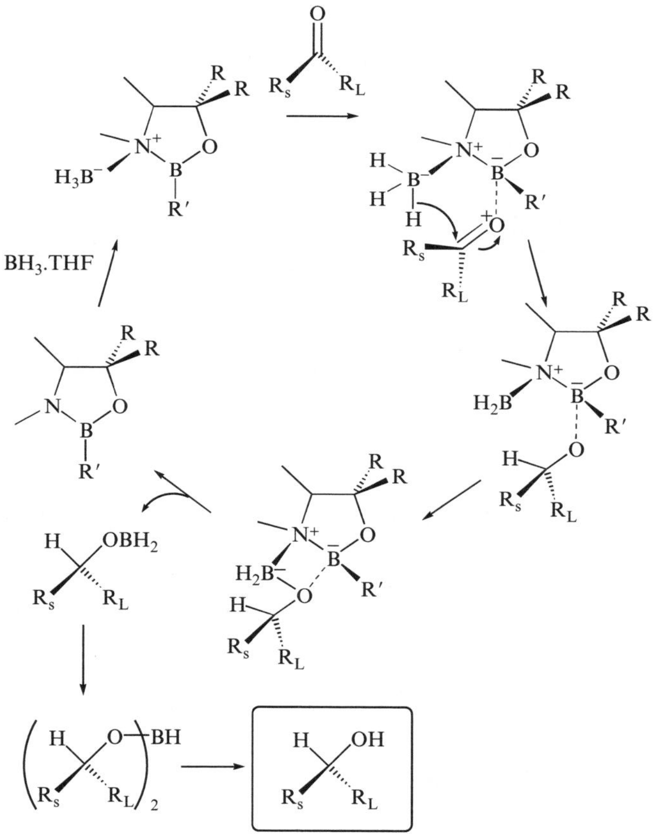

Figure 11.3   Mechanism of the reduction of ketone by borane catalysts[6].

11.2   OXAZABOROLIDINE BORANE REDUCTION OF ACETOPHENONE[7]

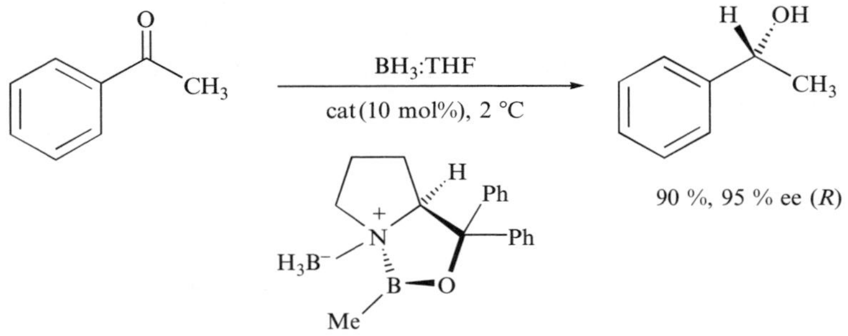

BH$_3$:THF

cat(10 mol%), 2 °C

90 %, 95 % ee (R)

Materials and equipment

- (S)-3,3-Diphenyl-1-methyltetrahydro-3H-pyrrolo[1,2-c][1,3,2]oxazaborole: (S)-Me-CBS, solution 1 M in toluene, 1.2 mL, 1.2 mmol, 0.12 eq
  The Me-CBS needs to be recently obtained and stored under argon; if a precipitate appears it can be due to the decomposition of the complex which is air and moisture sensitive.
- Anhydrous tetrahydrofuran, 15 mL
- Borane tetrahydrofuran complex, BH$_3$.THF, 1 M in tetrahydrofuran, 6.7 mL, 6.7 mmol, 0.67 eq.
  Borane complexes are water and air sensitive and need to be stored under argon in anhydrous conditions.
- Acetophenone, 1.2 g, 10 mmol
  Acetophenone was previously distilled under vacuum and stored under nitrogen.
- Aqueous solution of hydrochloric acid 1 N, 10 mL
- Petroleum ether, ethyl acetate, methanol, diethyl ether
- Brine
- Magnesium sulfate
- p-Anisaldehyde

- 100 mL Two-necked round-bottomed flask with a magnetic stirrer bar
- Magnetic stirrer hot plate with a thermostatically controlled oil bath and thermometer
- Addition funnel, 20 mL
- Ice-bath
- Separating funnel, 250 mL
- Rotary evaporator
- Kugelrohr apparatus

Procedure

1. A 100 mL two-necked round-bottomed flask equipped with a magnetic stirrer bar and an addition funnel were dried in an oven at 120 °C overnight. The dry flask, equipped with the addition funnel, was placed under vacuum until cool and then flushed with nitrogen.
2. The flask was charged with (S)-Me-CBS (1 M solution in toluene, 1.2 mL) in 10 mL of tetrahydrofuran. The mixture was cooled with an ice-bath and then BH$_3$.THF (6.7 mL) was added. The solution was stirred for 15 minutes.
3. The addition funnel was filled with acetophenone (1.16 mL) and dry tetrahydrofuran (5 mL); this solution was then added over 2 hours to the cold reaction mixture.
4. After completion, the reaction was stirred for an additional 30 minutes at room temperature.
5. The reaction was followed by TLC (eluent: petroleum ether–ethyl acetate; 75:25). The acetophenone was UV active, stained yellow with p-anisaldehyde, R_f 0.68. Phenylethanol had a low UV activity, stained purple with p-anisaldehyde, R_f 0.46.
6. The reaction was quenched by careful addition of methanol (5 mL, hydrogen evolution). An aqueous solution of hydrochloric acid 1N (10 mL) was then added and a white suspension appeared. The mixture was stirred for 15 minutes.
7. Diethyl ether was added (30 mL) and the two-phase solution was transferred into a separating funnel. The organic phase was separated and the aqueous layer extracted with diethyl ether (2 × 20 mL). The combined organic layers were washed with water (4 × 30 mL) and with brine (2 × 30 mL), dried over magnesium sulfate, filtered and concentrated to give a yellow oil (1.64 g).
8. The residue was purified by Kugelrohr distillation giving the phenylethanol as a colourless oil (1.1 g, 90 %).

   The ee (95 %) was determined by chiral GC (Lipodex® E, 25 m, 0.25 mm ID, temperatures: column 80 °C isotherm, injector 250 °C, detector 250 °C, mobile phase helium). R_t (S)-enantiomer: 68.3 min, R_t (R)-enantiomer: 71.1 min.

   ^{1}H NMR (200 MHz, CDCl$_3$): δ 7.18–7.36 (m, 5H, Ph); 4.87 (qd, J 6.6 Hz, J 3.3 Hz, 1H, CHOH); 2.25 (br s, 1H, OH); 1.48 (d, J 6.6 Hz, 3H, CH$_3$).

   IR (CHCl$_3$, cm^{-1}): 3611, 3458 (O–H), 3011, 2981 (C–H Ar), 2889 (C–H aliphatic), 1603 (Ar), 1493, 1453 (Ar), 1379 (Ar), 1255, 1075 (O–H), 895, 693 (Ar).

   Mass: calculated for C$_8$H$_{10}$O: m/z 122.07317, found [M]$^+$ 122.07293.

Conclusion

The reduction using oxazaborolidine borane needs to be done in anhydrous conditions to avoid the decomposition of the catalyst. The addition of acetophenone has to be as slow as possible to obtain a good enantiomeric excess. However, the reaction is easy to handle, the catalyst is commercially available

Table 11.1   Asymmetric reduction of ketones catalysed by (S)-Me-CBS[7] (results according to the literature).

	ee % (configuration)
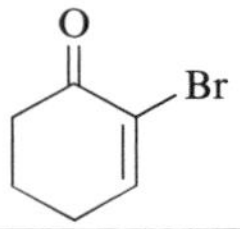	91 (R)
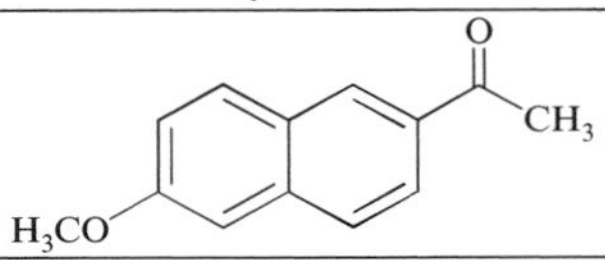	97.6 (R)
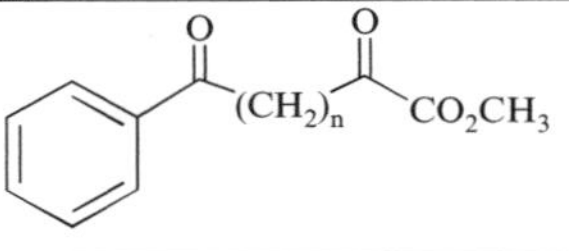	n = 2; 94 (R) n = 3; 96.7 (R)

although it has to be stored under argon to avoid decomposition. Table 11.1 gives some examples of the different substrates that can be reduced by oxazaborolidine borane complex, using the procedure described; other examples are given in Table 11.4. Some modifications of this method, using other hydrogen donor and/or other amino alcohols as catalyst ligands have been reported [6,8–12].

11.3   OXAZAPHOSPHINAMIDE BORANE REDUCTION OF CHLOROACETOPHENONE[13]

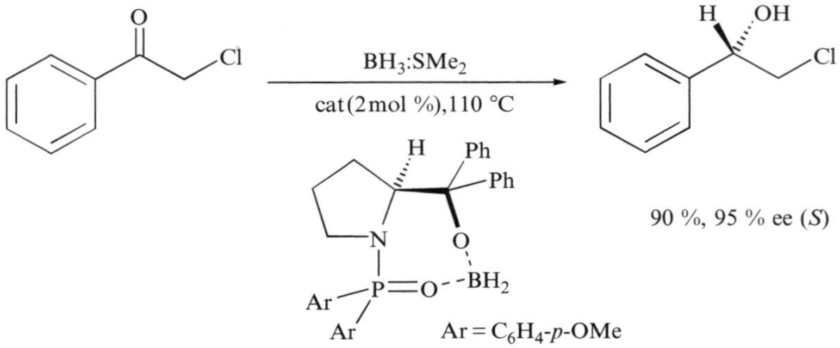

Materials and equipment

- 2-Chloroacetophenone, 154 mg, 1 mmol
- Anhydrous toluene, 16 mL

- Oxazaphosphinamide (N-(di-p-anisylphosphoryl)-(S)-α, α-diphenyl-2-pyrrolidine methanol), 50 mg, 0.1 mmol, 0.1 eq*
    The catalyst was prepared by reaction of (S)-diphenylprolinol with dimethylphosphinite and triethylamine in the presence of carbon tetrachloride. The N-(O,O-dimethylphosphoryl) derivative obtained was treated with an excess of p-anisylmagnesium bromide to give the oxazaphosphinamide catalyst[13].
- Borane dimethyl sulfide complex 2 M solution in tetrahydrofuran, 0.5 mL, 1 mmol, 1 eq
- Petroleum ether, ethyl acetate, triethylamine
- Saturated aqueous solution of NH_4Cl, 10 mL
- Brine
- Magnesium sulfate
- Silica gel 60 (0.063–0.04 mm)
- p-Anisaldehyde dip

- 50 mL Two-necked dry round-bottomed flask with a magnetic stirrer bar
- Magnetic stirrer hot plate with a thermostatically controlled oil bath and thermometer
- Dean and Stark apparatus
- Condenser
- Syringe, 3 mL
- Syringe pump
- Separating funnel, 250 mL
- Rotary evaporator
- Kugelrohr apparatus

Procedure

1. A 50 mL two-necked round-bottomed flask (dried overnight at 150 °C and cooled under vacuum) was equipped with a Dean and Stark apparatus and flushed with nitrogen.
2. The flask was filled with the catalyst (50 mg) and anhydrous toluene (4 mL). The mixture was refluxed until 3.5 mL of solvent was recovered. The catalyst was azeotroped twice with toluene (4 mL) and then cooled to room temperature under argon.
    Precautions were taken whilst azeotroping the catalyst with toluene: thus the use of freshly dried toluene and flame-dried glassware were necessary to ensure anhydrous conditions.
3. The Dean and Stark apparatus was removed, replaced by a condenser (the solution was flushed continuously with nitrogen) and the catalyst dissolved in anhydrous toluene (2 mL). Borane–dimethylsulfide (0.5 mL of a 2 M solution in tetrahydrofuran) was added to the mixture, which was heated to 110 °C.

---

* *The catalyst was kindly provided by Prof. M. Wills (University of Warwick, Coventry, UK).*

4. When the reaction was at reflux, a solution of chloroacetophenone (154 mg) in toluene (2 mL) was added via a syringe pump over 10 minutes. After completion of the addition the reaction was stirred for a further 20 minutes.
5. The reaction was followed by TLC (eluent: petroleum ether–ethyl acetate; 85:15). The chloroacetophenone was UV active and stained grey with p-anisaldehyde dip, R_f 0.5. 2-Chloro-1-phenylethanol was UV active and stained green-grey with p-anisaldehyde, R_f 0.39.
6. The mixture was cooled to room temperature and the borane–dimethylsulfide was slowly hydrolysed by water (10 mL) and then by a saturated solution of NH_4Cl (10 mL).
7. The mixture was transferred into a separating funnel and the two phases were separated. The aqueous layer was extracted with ethyl acetate (2 × 30 mL). The combined organic layers were washed with water (3 × 30 mL), brine (3 × 30 mL) and then dried over magnesium sulfate, filtered and concentrated to give a crude oil (620 mg).
8. The crude material was purified by flash chromatography on silica gel (30 g) using petroleum ether–ethyl acetate–triethylamine (89:10:1) as eluent to give 2-chloro-1-phenylethanol as an oil (140 mg, 90 %).

The ee (95 %) was determined by chiral GC (Lipodex® E, 25 m, 0.25 mm ID, temperatures: column 105 °C isotherm, injector 250 °C, detector 250 °C, mobile phase helium). R_t (R)-enantiomer: 102.4 min; R_t (S)-enantiomer: 106.7 min.

^{1}H NMR (200 MHz, CDCl$_3$): δ 7.39–7.31 (m, 5H, Ph); 4.88 (ddd, J 8.8 Hz, J 3.3 Hz, J 3.3 Hz, 1H, CH); 3.74 (dd, J 3.3 Hz, J 11.5 Hz, 1H, CH_aH$_b$); 3.70 (dd, J 8.8 Hz, J 11.8 Hz, 1H, CH$_a$$H_b$); 2.78 (br s, 1H, O$H$).

IR (CHCl$_3$, cm^{-1}): 3586, 3460 (O–H), 3070, 3012, (C–H Ar), 2961, 2897 (C–H aliphatic), 1603 (Ar), 1494, 1454 (Ar), 1428, 1385, 1254, 1187, 1062, 1012, 870, 690.

Mass: calculated for C_8H_9OCl: m/z 156.03419, found [M]$^+$ 156.03385.

Conclusion

The reduction using the oxazaphosphinamide is easy to reproduce and the results correlate with the published material. During the reaction the addition of the chloroacetophenone solution needs to be as slow as possible; this is an essential factor for obtaining a good enantiomeric excess. According to the publication, the reaction could be performed without the prescribed precautions to work under anhydrous conditions with only a small drop in selectivity and no change to the reaction time. This is due to the stability of the phosphinamide reagent, which is not sensitive to water or oxygen. Another advantage of using this catalyst is that it does not decompose under the reaction conditions and could be recovered and re-used without any decrease in the reactivity. In Table 11.2 different results obtained by oxazaphosphinamide catalysts are reported. Some other examples are given in Table 11.4.

Table 11.2 Reduction of aromatic ketones using oxazaphosphinamide catalyst[13] (results according to the literature).

	Yield %	ee %
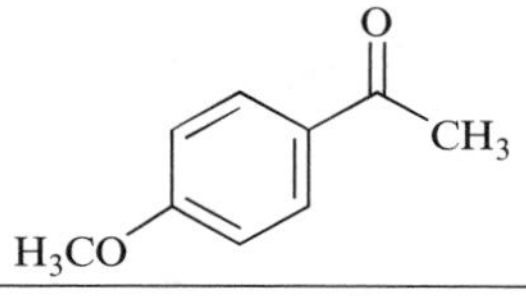	89	90
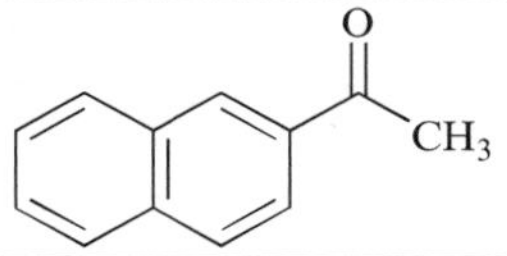	82	90
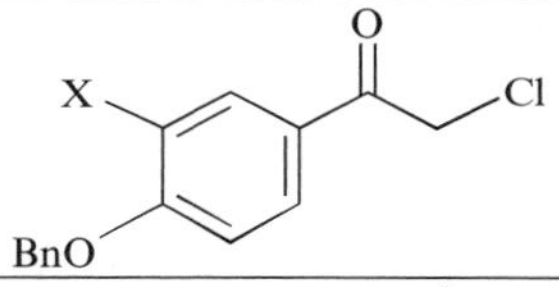	X = H; 8 X = CH$_2$OBn; 84	94 93
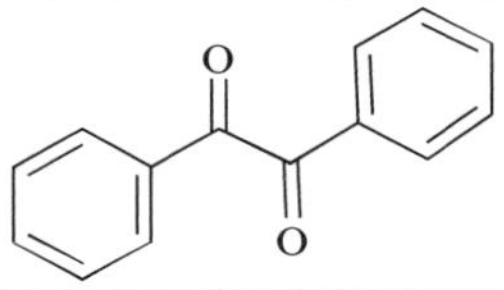	83	>90
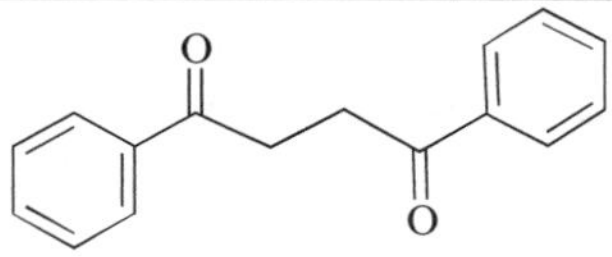	71	>90

11.4 ASYMMETRIC REDUCTION OF CHLOROACETOPHENONE USING A SULFOXIMINE CATALYST [S]

11.4.1 PREPARATION OF β-HYDROXYSULFOXIMINE BORANE

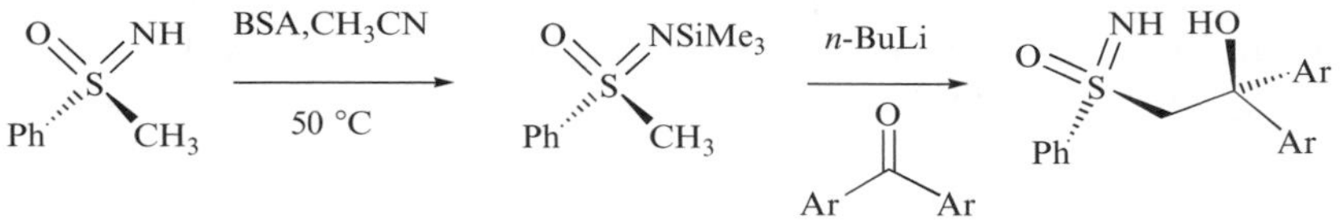

Materials and equipment

- (SS)-Methyl-S-phenylsulfoximine, 523 mg, 3.4 mmol*

* *(SS)-Methyl-S-phenylsulfoximine was kindly provided by Prof. C. Bolm (Technische RWTH Aachen)*

- *N, O*-Bis-(trimethylsilyl)-acetamide (BSA), 924 μL, 3.74 mmol, 1.1 eq
- Dry acetonitrile, 15 mL
- Dry tetrahydrofuran, 18 mL
- *n*-Butyl lithium, 1.6 *M* in hexane, 2.1 mL, 3.4 mmol, 1 eq
- Benzophenone, 682 mg, 3.74 mmol, 1.1 eq
- Aqueous saturated solution of NH_4Cl and methanol (10:1), 2 mL
- Petroleum ether, ethyl acetate
- *p*-Anisaldehyde dip
- Silica gel 60 (0.063–0.04 mm)

- 50 mL Two-necked round-bottomed flask with a magnetic stirrer bar
- 50 mL Schlenk tube with a magnetic stirrer bar
- Condenser
- Cannula (double-tipped needle)
- Magnetic stirrer hot plate with a thermostatically controlled oil bath and thermometer
- Ice-bath
- Syringe
- Solid carbon dioxide/ethanol cooling bath (−78 °C)
- Kugelrohr apparatus

Procedure

1. A 50 mL two-necked flask equipped with a magnetic stirrer bar was dried overnight at 150 °C, cooled under vacuum and flushed with nitrogen.
2. Under a nitrogen atmosphere, the flask was charged with (*SS*)-methyl-S-phenylsulfoximine (523 mg) and placed under vacuum. The flask was flushed with nitrogen, then dry acetonitrile (15 mL), and *N, O*-bis-(trimethylsilyl)-acetamide (924 μL) were added.
3. The flask was equipped with a condenser and the mixture was heated under nitrogen at 50 °C and stirred for 45 minutes at this temperature (it was not necessary for the solvent to reflux).
4. After 45 minutes, the mixture was cooled to room temperature under nitrogen. The solvent was evaporated under high vacuum.
5. The residue was placed under nitrogen in a Kugelrhor apparatus and the impurities were distilled at a temperature less than 70 °C at 0.3 mbar. The purity of N-(trimethylsilyl)-S-methyl-S-phenylsulfoximine was verified by NMR.
   - ^{1}H NMR(200 MHz, CDCl$_3$): δ 7.97–7.93 (m, 2H, Ph); 7.57–7.451 (m, 3H, Ph); 3.01 (s, 3H, C*H*$_3$); 0.11 (s, 9H, Si (C*H*$_3$)$_3$).
6. A 50 mL Schlenk tube equipped with a magnetic stirrer bar, dried overnight at 150 °C, was cooled under vacuum and then flushed with nitrogen.
7. N-(Trimethylsilyl)-S-methyl-S-phenylsulfoximine (prepared as above) was dissolved in 15 mL of dry tetrahydrofuran and transferred by cannula into the Schlenk tube under nitrogen.

8. The mixture was cooled to 0 °C with an ice-bath and then 2.1 mL of *n*-BuLi (1.6 *M* in hexane) was added carefully via a syringe. The solution became yellow.

9. The mixture was cooled to −78 °C using an ethanol cooling bath. A solution of benzophenone (682 mg) in dry tetrahydrofuran (3 mL) was then added dropwise. The mixture was stirred for 2 hours at −78 °C.

10. The reaction was followed by TLC (eluent: petroleum ether–ethyl acetate; 9:1). The benzophenone was UV active and stained yellow with permanganate, R_f 0.58. β-Hydroxysulphoximine was UV active and stained yellow with *p*-anisaldehyde, R_f 0.11.

11. The reaction was quenched with aqueous saturated solution of NH_4Cl and methanol (10:1, 2 mL). The mixture was stirred overnight at room temperature and the solvent was evaporated under reduced pressure.

12. The alcohol was obtained by flash chromatography on silica gel eluting with petroleum ether–ethyl acetate (9:1) to eliminate the benzophenone and then with an eluant ratio 6:4, giving (*SS*)-1,1-diphenyl-2-(S-phenylsulfonimidoyl)-ethanol (790 mg, 2.3 mmol, 77%).

- The yield of the reaction is variable (33–77%), especially if the reaction is not carried out under strictly anhydrous conditions or if the flash chromatography takes an excessive amount of time.
- ^{1}H NMR(200 MHz, CDCl$_3$): δ 7.6–7.09 (m, 15H, Ph); 4.11 (s, 2H, CH$_2$); 2.96 (br s, 1H, NH).

11.4.2 REDUCTION OF CHLOROACETOPHENONE USING THE SULFOXIMINE BORANE

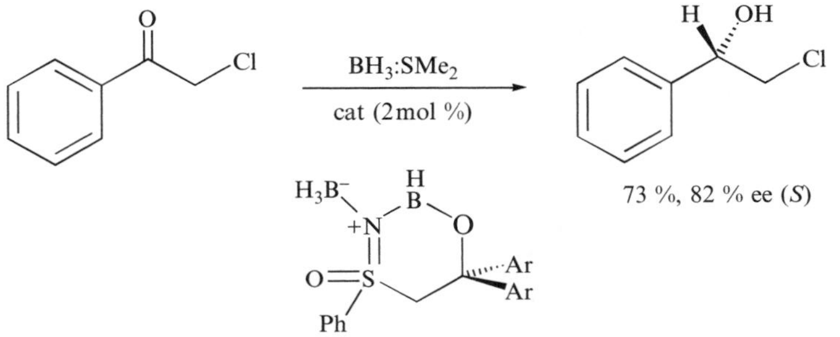

Materials and equipment

- Sulfoximine catalyst, (*SS*)-1, 1-diphenyl-2-(S-phenylsulfonimidoyl)-ethanol, 68 mg, 0.2 mmol, 0.1 eq
- Dry toluene, 6 mL
  The toluene was distilled from sodium and benzophenone and then stored over activated molecular sieves.

- Borane dimethylsulfide, $2\,M$ in THF, 1.2 mL, 2.4 mmol, 1.2 eq
- Chloroacetophenone, 310 mg, 2 mmol

  Chloroacetophenone is toxic and needs to be manipulated using gloves and eye protection in a well-ventilated fume-hood.

- Aqueous solution of hydrochloric acid, 1 N, 3 mL
- Diethyl ether
- Aqueous solution of sodium hydroxide, 2 N, 20 mL
- Sodium sulfate
- *p*-Anisaldehyde dip

- 50 mL Two-necked round-bottomed flask with a magnetic stirrer bar
- Magnetic stirrer hot plate with a thermostatically controlled oil bath and thermometer
- Syringe pump
- Syringe, 3 mL
- Separating funnel, 250 mL
- Kugelrohr apparatus

Procedure

1. A 50 mL round-bottomed flask equipped with a magnetic stirrer was dried overnight at 150 °C and placed under vacuum and then flushed with nitrogen.
2. The flask was charged with the sulfoximine catalyst (68 mg) and dry toluene (4 mL). To this white suspension was added borane dimethylsulfide (1.2 mL). The mixture became clear with the evolution of hydrogen.
3. After 15 minutes a solution of chloroacetophenone (310 mg) in dry toluene (2 mL) was added via a syringe pump over a period of 3 hours at room temperature.
4. After completion of the addition the mixture was stirred for a further 10 minutes. The reaction was quenched with an aqueous solution of HCl (1 N, 3 mL) and water (10 mL).
5. The mixture was transferred into a separating funnel and the two phases separated. The aqueous layer was extracted with diethyl ether (3×30 mL) and the combined organic layers washed with an aqueous solution of sodium hydroxide (2 N, 20 mL) and then dried over sodium sulfate, filtered and concentrated.
6. The alcohol was obtained by distillation of the residue using a Kugelrohr apparatus (120 °C, 3 mmHg) to give (S)-2-chloro-1-phenylethanol (233 mg, 1.49 mmol, 73 %).

   The ee (82 %) was determined by chiral GC analysis (Lipodex® E, 25 m, 0.25 mm ID, temperatures: column 120 °C isotherm, injector 250 °C, detector 250 °C, mobile phase helium) R_t (R)-enantiomer: 45.3 min, R_t (S)-enantiomer: 46.5 min.

^{1}H NMR(200 MHz, CDCl$_3$): δ 7.39–7.31 (m, 5H, Ph); 4.88 (ddd, J 8.8 Hz, J 3.3 Hz, J 3.3 Hz, 1H, CH); 3.74 (dd, J 3.3 Hz, J 11.5 Hz, 1H, CH_aH$_b$); 3.70 (dd, J 8.8 Hz, J 11.8 Hz, 1H, CH$_a$$H_b$); 2.78 (br s, 1H, O$H$).

Conclusion

To obtain a good enantiomeric excess, the ligand synthesis and the reduction reaction need to be carried out under strictly anhydrous conditions. The addition of the substrate needs to be as slow as possible. Table 11.3 gives some examples of the different substrates that can be reduced by the hydroxysulfoximine-borane catalyst described. Other examples are given in the comparative Table 11.4. Concerning the synthesis of the catalyst, the yield can dramatically decrease if the reaction conditions are not strictly anhydrous.

Table 11.3   Reduction of ketones by hydroxysulfoximine-borane catalyst[14] (results according to the literature).

	ee % (configuration)
(Ph–CH$_2$CH$_2$–C(=O)–CH$_3$)	70 (R)
(Ph–C(=O)–CH$_2$–Br)	81 (S)
(Ph–C(=O)–CH$_2$–O-DMTr)	93 (S)
(Ph–C(=O)–CH$_2$–OSiPh$_2$-tBu)	92 (S)

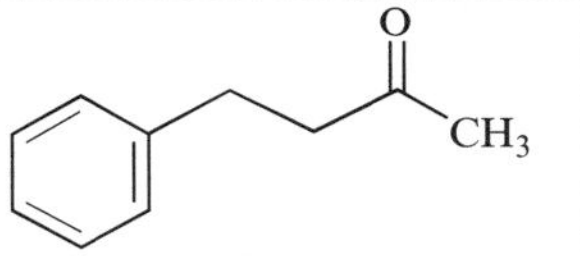

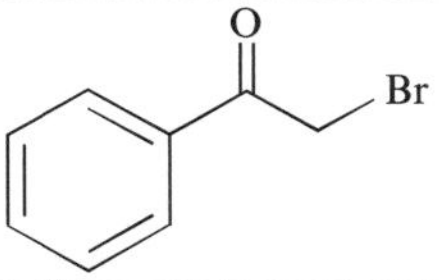

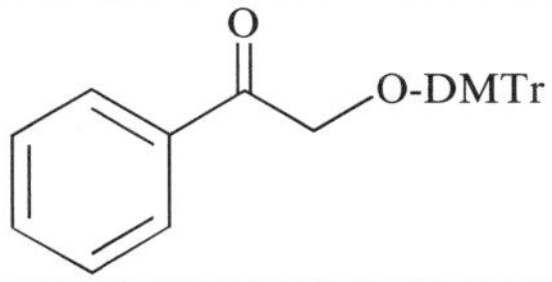

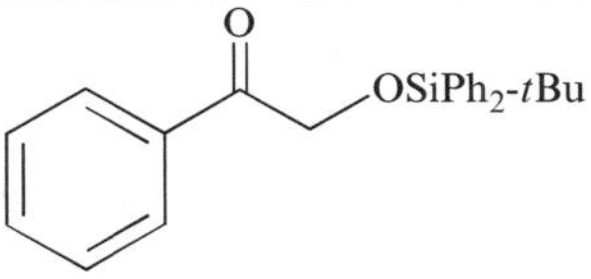

11.4.3   SUMMARY

All the different methods using nonmetallic catalysts are similar in terms of procedure; they all require anhydrous conditions to obtain high enantiomeric excesses. However, the oxazaphosphinamide catalysts can give relatively high enantiomeric excess without all the precautions of reactions conducted under strictly anhydrous conditions. Table 11.4 gives some substrates that can be reduced by the three catalysts described above. Each catalyst can give good results depending on the nature of the substrate. However, considering the results and the commercially availability, the reduction of ketones with Corey's catalyst is the easiest method to use.

Table 11.4 Catalytic reduction of ketones by nonmetallic catalysts (results according to the relevant publications).

	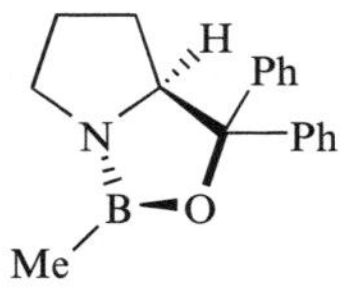 Oxaza borolidines[7] ee % (configuration)	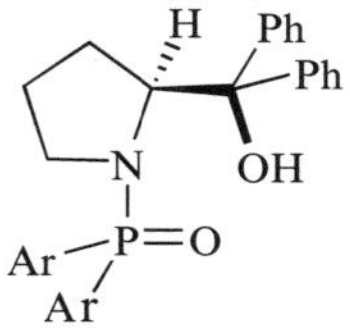 Oxaza phosphinamides[13] Yield %, ee % (configuration)	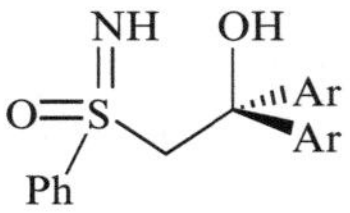 Hydroxy sulfoximines[14] ee % (configuration)
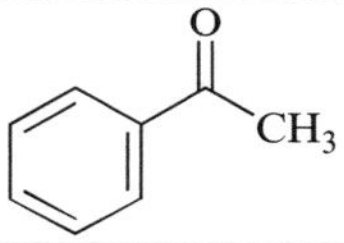 Ph-C(=O)-CH₃	96.5 (R)*	83, 88 (R)	76 (R)
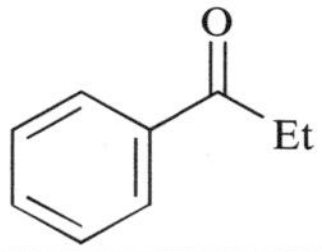 Ph-C(=O)-Et	96.7 (R)	76, 77 (R)	73 (R)
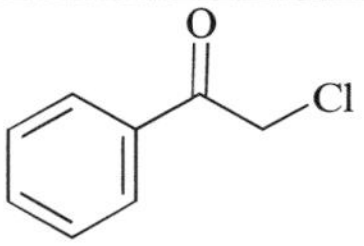 Ph-C(=O)-CH₂Cl	95.3 (S)	91, 94.4 (S)*	84 (S)*
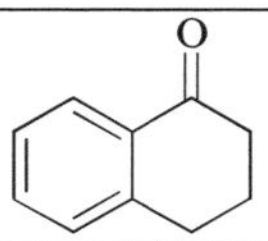 1-tetralone	86 (R)	81, 82 (R)	–
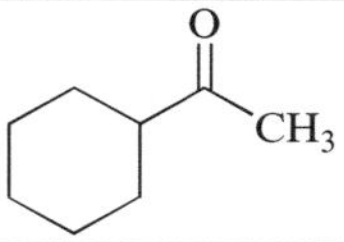 cyclohexyl-C(=O)-CH₃	84 (R)	87, 67 (R)	–
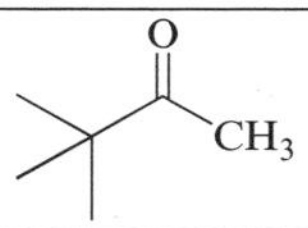 tBu-C(=O)-CH₃	97.3 (R)	65-86 (R)	–

* Reaction validated

11.5   ASYMMETRIC REDUCTION OF BROMOKETONE CATALYZED BY *CIS*-AMINOINDANOL OXAZABOROLIDINE

CHRIS H. SENANAYAKE, H. SCOTT WILKINSON and GERALD J. TANOURY

Chemical Research and Development, Sepracor Inc., 111 Locke Drive, Marlbrough, MA 01752, USA

11.5.1   SYNTHESIS OF AMINOINDANOL OXAZABOROLIDINE[15]

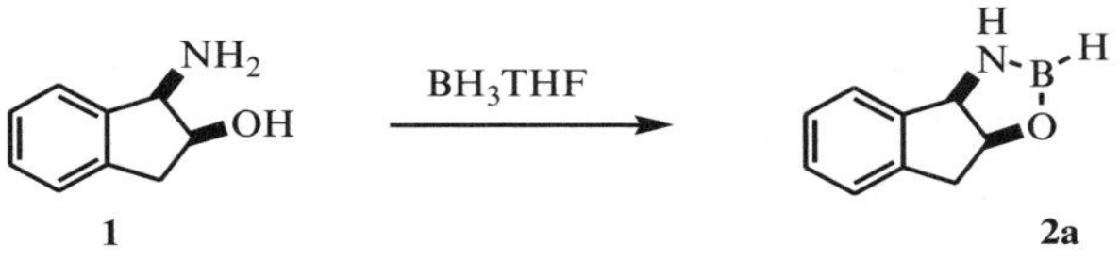

Materials and equipment

- (1*R*,2*S*)-Aminoindanol, 2.0 g
- Anhydrous tetrahydrofuran, 100 mL
- Borane-THF (1.0 M), 290 mL

- 2000 mL Round-bottomed flask with an overhead stirrer
- Mechanical stirrer

Procedure

1. A 2 L dried round-bottomed flask under an inert atmosphere was charged with aminoindanol (2.0 g) and anhydrous tetrahydrofuran (100 mL).
2. Borane–THF (1.0 M, 290 mL) was added while the temperature was maintained between 0–25 °C.
3. The mixture was stirred for 30 minutes at 20 °C.

11.5.2   ASYMMETRIC REDUCTION OF 2-BROMO-(3-NITRO-4-BENZYLOXY)ACETOPHENONE[16]

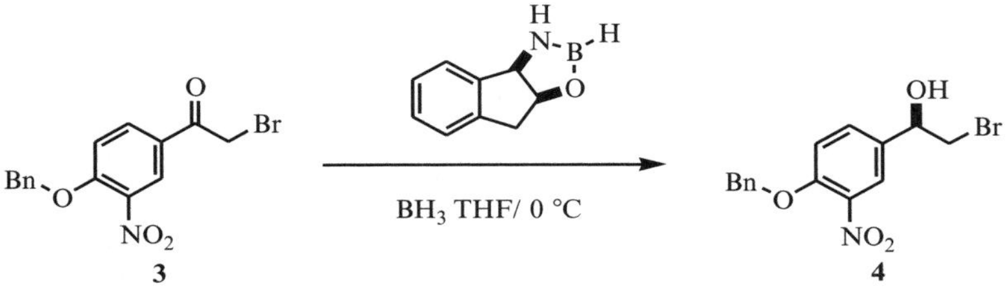

Materials and equipment

- 2-Bromo-(3-nitro-4-benzyloxy)acetophenone, 100 g
- Anhydrous tetrahydrofuran, 800 mL
- (1R,2S)-Aminoindanol-oxazaborolidine
- Acetone, 100 mL
- Toluene, 700 mL
- Aqueous 2 % sulfuric acid solution, 350 mL
- Aqueous 20 % NaCl solution, 160 mL
- Heptane, 200 mL
- Heptane, 200 mL

- 1 L Round-bottomed flask with a magnetic stirrer bar
- Magnetic stirrer
- Separatory funnel (2 L)
- Rotary evaporator
- Buchner funnel

Procedure

1. The oxazaborolidine solution was cooled to 0 °C.
2. A solution of bromoketone **3** in tetrahydrofuran (100 g in 800 mL THF, 0.35 M) was slowly added over 1 hour to the oxazoborolidine solution while the temperature was maintained between 0–5 °C. The mixture was stirred for 30 minutes at 0 °C.
3. Acetone (100 mL) was slowly added to quench the excess borane. The reaction mixture was concentrated to 300 mL and toluene (700 mL) was added. The solution was washed with 2 % sulfuric acid (350 g) then with 20 % NaCl (120 g). The organic phase was concentrated to 300 mL and cooled to 5 °C.
4. The resulting slurry was stirred at 5 °C for 1 hour, heptane (200 mL) was slowlyadded and the mixture was stirred an additional 1 hour at 5 °C.
5. The slurry was filtered and the solid was washed with heptane (200 mL). The off-white solid was dried *in vacuo* to give 89 g (93 % ee, >99 % cp) of the desired alcohol.

Recrystallization of 2-bromo-(3-nitro-4-benzyloxyphenyl)ethanol

Materials and equipment

- 2-Bromo-(3-nitro-4-benzyloxy)acetophenone, 48 g
- Toluene, 100 mL
- Heptane, 125 mL

- 500 mL Round-bottomed flask with a magnetic stirrer bar

- Magnetic stirrer
- Buchner funnel

Procedure

1. 2-Bromo-(3-nitro-4-benzyloxyphenyl) ethanol (93 % ee, 48 g) and 100 mL of toluene were placed in a 500 mL flask. The mixture was warmed until all the alcohol dissolved. The mixture was cooled to 5 °C and stirred for 1 hour.
2. Heptane (100 mL) was slowly added to the stirring mixture and that solution was stirred for 1 hour at 5 °C.
3. The slurry was filtered and the solid washed with heptane (25 mL).
4. The solid was dried in a vacuum oven to yield 45 g of 2-bromo-(3-nitro-4-benzyloxyphenyl) ethanol (>99 % ee).

    ^{1}H NMR (300 MHz DMSO-d6): δ 7.88 (m, 1H), 7.65 (d, 1H), 7.3–7.5 (m, 6 H), 6.01 (d, 1H), 5.35 (s, 2 H), 4.83 (m, 1H), 3.64 (ddd, 2H).

    ^{13}C NMR (MHz DMSO-d6): δ 151.94, 140.15, 135.52, 133.24, 131.73, 128.95, 128.51, 127.51, 123.62, 115.47, 72.40, 71.45, 39.75.

    IR: (KBr): 3381 (OH), 3091, 3067, 2961, 2893 (C–H), 1532, 1296, 1026, 729 cm^{-1}.

11.5.3   CONCLUSIONS

This procedure has been developed through the evaluation of several reaction parameters (catalyst, temperature, borane source, additives) and has been successfully used on large scale. The chemical purity of the product is excellent and the enantiomeric purity of the product can be increased by crystallizing from toluene/heptane.

The temperature has a significant effect on the selectivity of the reaction, with the optimal temperature being dependent on the borane source. The optimal range of temperature was 25 °C when borane–dimethylsulfide was used and 0–5 °C when borane–tetrahydrofuran was used as the reducing agent (Table 11.5).

Table 11.5 Optimization of enantioselectivity as a function of borane source and temperature using aminoindanol oxazaborolidine.

Borane	Temperature	er (% R)	ee (%)
	40	89.0	78
BH$_3$ − Me$_2$S	**25**	**95.0**	**90**
	0	91.0	82
	−10	66.0	32
	25	95	90
BH$_3$ − THF	**0**	**96.5**	**93**
	−10	94	88

Table 11.6 Effect of catalyst ratio and additives on % ee.

Entry	Mol % Catalyst	Additive	% ee
1	1	None	88
2	5	None	93
3	10	None	93
4	10	H_2O (5)	85
5	10	H_2O (20)	50
6	10	CH_3CN (20)	92
7	10	2-propanol (20)	91

The minimum amount of catalyst needed to obtain maximum selectivity was determined to be 5 mol%. Larger quantities had no effect. Consistent with other literature reports[17], very small quantities of water (5 mol% = 2.5 mg H_2O/g **3**) lowered the selectivities (Table 11.6, entry 4). Water sensitivity required thorough drying of the equipment, the starting materials and the solvents. In the case of tetrahydrofuran, drying was achieved by using activated 5 Å molecular sieves (KF titration >0.005%). On the other hand, solvents used for crystallization of the starting material (**3**), such as 2-propanol and acetonitrile showed little effect on the enantioselectivities of the reaction (entries 6 and 7).

After finding the optimal condition for catalyst **2a** in the reduction process, studies were aimed at understanding the role of the rigid indane platform, which behaves as a conformationally restricted phenyl glycinol equivalent. The use of the homologous six-membered[18] catalyst **5** in the asymmetric reduction process was examined. Surprisingly, the less rigid B–H catalyst **5a** displayed a higher degree of enantioselection than the corresponding indane catalyst **2a** (Table 11.6), while B–Me catalyst **5b** displayed similar selectivity compared to B–Me catalyst **2b**. The increased selectivity of catalyst **5a** may be due to the closer proximity of the C_{ortho} – H to the N–BH_3 moiety when compared to catalyst **2a**.

This study has clearly shown that B–H and B–Me catalysts have different optimal conditions for each catalyst system in the reduction of prochiral

Table 11.7 Comparison of rigid aminoalcohols and catalyst types (B-methyl vs. B–H).

Catalyst b R = H	(R)-**5a**, 96 % ee, 0 °C, BH_3•THF	(R)-**2a**, 93 % ee, 0 °C, BH_3•THF	(R)-**6a**, 26 % ee, 25 °C, BMS*
Catalyst a R = Me[20]	(R)-**5b**, 95 % ee –10 °C, BMS	(R)-**2b**, 96 % ee, 0 °C, BH_3•THF	(R)-**6b**, 12 % ee, –10 °C, BH_3•THF*

*Unoptimized

ketones. The highest selectivities are observed with catalyst **5a** (tetralin platform) and catalyst **2b**, and the lowest with catalysts **6a** and **6b**. From a practical point of view, B–H catalyst systems are much more preferred than B–alkyl systems. Therefore, the use of highly effective B–H oxazaborolidine catalysts from readily accessible tetralin and aminoindanol is recommended.

11.5.4   STEREOSELECTIVE REDUCTION OF 2,3-BUTADIONE MONOXIME TRITYL ETHER

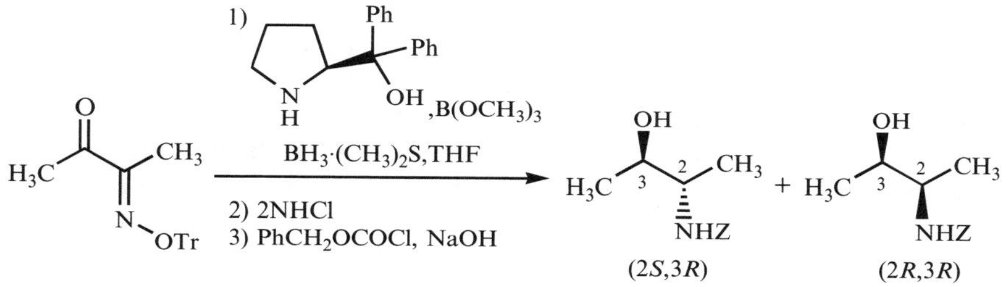

Materials and equipment

- Anhydrous tetrahydrofuran, 10 mL
- 2,3-Butadionemonoxime trityl ether, 1.72 g, 5.0 mmol
- (S)-α, α-Diphenylpyrrolidinemethanol, 127 mg, 0.5 mmol
- Trimethyl borate, 62 mg, 0.6 mmol
- 10 M Borane–dimethylsulfide complex, 2.0 mL, 20 mmol
- 2 N Hydrochloric acid, 15 mL, 30 mmol
- Sodium hydroxide, 2.4 g, 60 mmol
- Benzyl chloroformate, 3.41 g, 20 mmol
- Diethyl ether, 30 mL
- Methylene chloride, 60 mL
- Magnesium sulfate
- Silica gel
- n-Hexane, ethyl acetate

- 25 mL Three-necked flask with a magnetic stirrer bar
- 200 mL Round-bottomed flask with a magnetic stirrer bar
- Magnetic stirrer
- Ice-bath
- Oil-bath
- Separating funnel, 100 mL
- Rotary evaporator

Procedure

1. (S)-α, α-Diphenylpyrrolidinemethanol (127 mg) was placed in a 25 mL three-necked flask equipped with a magnetic stirrer bar, under nitrogen. A solution of trimethyl borate (62 mg) in dry tetrahydrofuran (5 mL) was added. The mixture was stirred for 1 hour at room temperature.
2. 10 M Borane–dimethylsulfide complex (2.0 mL) was added to the resulting solution. The mixture was cooled to 0–5 °C with an ice-bath, and then a solution of 2,3-butadione monoxime trityl ether (1.72 g) in dry tetrahydrofuran (5 mL) was added dropwise via a syringe pump over 1 hour at that temperature.
3. After being stirred for 0.5 hour at 0–5 °C, the mixture was allowed to warm to room temperature and heated under reflux for 18 hours. The resulting mixture was cooled to room temperature and cautiously transferred into 2 N hydrochloric acid (15 mL) in a 200 mL round-bottomed flask equipped with a magnetic stirrer bar using diethyl ether (10 mL).
4. After being stirred for 5 hours at room temperature, the mixture was made basic with sodium hydroxide (2.4 g). The organic solvents were removed under reduced pressure using a rotary evaporator. The aqueous residue was washed with diethyl ether (2 × 10 mL) and then benzyl chloroformate (3.41 g) was added. The mixture was stirred for 20 hours at room temperature.
5. The resulting mixture was transferred into a separating funnel with methylene chloride (20 mL) and the phases were separated. The aqueous layer was extracted with methylene chloride (2 × 20 mL). The combined organic layers were dried over magnesium sulfate, filtered and concentrated using a rotary evaporator.
6. The residue was purified by silica gel column chromatography using n-hexane–ethyl acetate (3:1 → 1:1) as an eluent to give as a white solid 3-benzyloxyamino-2-butanol (1.03 g, 92%) as a mixture of diastereomers.

   The anti/syn ratio (86:14) and the respective ee (anti 99%, syn 97%) were determined by HPLC (Chiralcel OJ chiral column (i.d. 4.6 × 250 mm), flow 0.5 mL/min, eluent n-hexane–isopropanol 9:1, detection UV 230 nm); 22.9 min for (2S, 3S)-isomer, 26.8 min for (2S, 3R)-isomer, 29.8 min for (2R, 3R)-isomer, 36.1 min for (2R, 3S)-isomer.

   ^{1}H NMR (270 MHz, CDCl$_3$) for anti isomer δ 1.11 (d, J 6.7 Hz, 3H), 1.15 (d, J 6.7 Hz, 3H), 2.18 (br, 1H), 3.74 (m, 1H), 3.88 (m, 1H), 4.94 (br, 1H), 5.10 (s, 2H), 7.35 (m, 5H); for syn isomer δ 1.18 (d, J 6.7 Hz, 3H), 1.20 (d, J 6.1 Hz, 3H), 1.88 (br, 1H), 3.70 (m, 2H), 4.94 (br, 1H), 5.10 (s, 2H), 7.35 (m, 5H).

11.5.5   STEREOSELECTIVE REDUCTION OF METHYL 3-OXO-2-TRITYLOXYIMINOSTEARATE

Materials and equipment

- Anhydrous tetrahydrofuran, 10 mL
- Methyl 3-oxo-2-trityloxyiminostearate, 1.46 g, 2.5 mmol
- (S)-α,α-Diphenylpyrrolidinemethanol, 63 mg, 0.25 mmol
- Trimethyl borate, 31 mg, 0.3 mmol
- Borane-diethylaniline complex, 815 mg, 5.0 mmol
- 10 M Borane–dimethylsulfide complex, 2.0 mL, 20 mmol
- 2 N Hydrochloric acid, 10 mL, 20 mmol
- Sodium hydroxide, 1.4 g, 35 mmol
- Benzoyl chloride, 0.70 g, 5.0 mmol
- Diethyl ether, methanol, methylene chloride, tetrahydrofuran
- Magnesium sulfate
- Silica gel

- 25 mL Three-necked flask with a magnetic stirrer bar
- 200 mL Round-bottomed flask with a magnetic stirrer bar
- Magnetic stirrer
- Oil-bath
- Separating funnel, 100 mL
- Rotary evaporator

Procedure

1. (S)-α,α-Diphenylpyrrolidinemethanol (63 mg) was placed in a 25 mL three-necked flask equipped with a magnetic stirrer bar, under nitrogen. A solution of trimethyl borate (31 mg) in dry tetrahydrofuran (5 mL) was added. The mixture was stirred for 1 hour at room temperature.
2. Borane–diethylaniline complex (815 mg) was added to the resulting mixture. A solution of methyl 3-oxo-2-trityloxyiminostearate (1.46 g) in dry tetrahydrofuran (5 mL) was added dropwise via a syringe pump over 1 hour at room temperature.

3. After being stirred for 2 hours at room temperature, 10 M borane–dimethyl-sulfide complex (2.0 mL) was added. The mixture was heated under reflux for 65 hours. The resulting mixture was cooled to room temperature and cautiously transferred into 2 N hydrochloric acid (10 mL) in a 200 mL round-bottomed flask equipped with a magnetic stirrer bar using diethyl ether (10 mL).

4. After being stirred for 1 hour at 60 °C, the mixture was cooled to room temperature and then made basic with sodium hydroxide (1.4 g). Benzoyl chloride (0.70 g) was added, and the mixture was stirred for 1 hour at room temperature and subsequently stirred for 2 hour at 60 °C with methanol (10 mL).

5. The organic solvents were removed under reduced pressure using a rotary evaporator. The residue was transferred into a separating funnel with methylene chloride–tetrahydrofuran (2:1, 40 mL) and the phases were separated. The aqueous layer was extracted with methylene chloride–tetrahydrofuran (2:1, 2 × 40 mL). The combined organic layers were dried over magnesium sulfate, filtered and concentrated using a rotary evaporator.

6. The residue was purified by silica gel column chromatography using methylene chloride–methanol (50:1 → 30:1) as an eluent to give a white solid *N*-benzoylsphingamine (0.93 g, 92%) as a mixture of diastereomers.

The *anti/syn* ratio (13:87) and the respective ee (*anti* 75%, *syn* 89%) were determined by HPLC (YMC Chiral NEA® chiral column (i.d. 4.6 × 250 mm) and Chiralcel OJ-R chiral column (i.d. 4.6 × 150 mm) connected in series, flow 0.3 mL/min, eluent acetonitrile–water 3:7, detection UV 254 nm); 65.6 min for (2*S*, 3*R*)-isomer, 68.0 min for (2*R*, 3*S*)-isomer, 71.8 min for (2*R*,3*R*)-isomer, 74.2 min for (2*S*,3*S*)-isomer.

^{1}H NMR (270 MHz, CDCl$_3$) for *syn* isomer δ 0.88 (t, *J* 6.7 Hz, 3H), 1.25 (br, 26H), 1.50 (m, 2H), 3.32 (br, 2H), , 3.88 (m, 2H), 4.03–4.14 (m, 2H), 6.97 (d, *J* 8.6 Hz, 1H), 7.40 (t, *J* 7.3 Hz, 2H), 7.48 (d, *J* 7.3 Hz, 1H), 7.78 (d, *J* 7.3 Hz, 2H).

11.5.6 STEREOSELECTIVE REDUCTION OF 1-(TERT-BUTYLDIMETHYLSILYLOXY)-3-OXO-2-TRITYLOXYIMINOOCTADECANE

Materials and equipment

- Anhydrous tetrahydrofuran, 10 mL
- 1-(*tert*-Butyldimethylsilyloxy)-3-oxo-2-trityloxyiminooctadecane, 1.68 g, 2.5 mmol
- (*S*)-α,α-Diphenylpyrrolidinemethanol, 63 mg, 0.25 mmol
- Trimethyl borate, 31 mg, 0.3 mmol
- Borane–diethylaniline complex, 815 mg, 5.0 mmol
- 10 M Borane–dimethylsulfide complex, 0.5 mL, 5.0 mmol
- 2 N Hydrochloric acid, 10 mL, 20 mmol
- Sodium hydroxide, 1.4 g, 35 mmol
- Benzoyl chloride, 0.70 g, 5.0 mmol
- Diethyl ether, methanol, methylene chloride, tetrahydrofuran
- Magnesium sulfate
- Silica gel

- 25 mL three-necked flask with a magnetic stirrer bar
- 200 mL round-bottomed flask with a magnetic stirrer bar
- Magnetic stirrer plate
- Oil-bath
- Separating funnel, 100 mL
- Rotary evaporator

Procedure

1. (*S*)-α,α-Diphenylpyrrolidinemethanol (63 mg) was placed in a 25 mL three-necked flask equipped with a magnetic stirrer bar, under nitrogen. A solution of trimethyl borate (31 mg) in dry tetrahydrofuran (5 mL) was added. The mixture was stirred for 1 hour at room temperature.
2. Borane–diethylaniline complex (815 mg) was added to the resulting mixture. A solution of 1-(*tert*-butyldimethylsilyloxy)-3-oxo-2-trityloxyiminooctadecane (1.68 g) in dry tetrahydrofuran (5 mL) was added dropwise using a syringe pump over 1 hour at room temperature.
3. After being stirred for 1 hour at room temperature, 10 M borane–dimethylsulfide complex (0.5 mL) was added. The mixture was heated under reflux for 18 hours. The resulting mixture was cooled to room temperature and cautiously transferred into 2 N hydrochloric acid (10 mL) in a 200 mL round-bottomed flask equipped with a magnetic stirrer bar using diethyl ether (10 mL).
4. The same procedure described for the stereoselective reduction of methyl 3-oxo-2-trityloxyiminostearate gave a white solid *N*-benzoylsphingamine (0.96 g, 94%) as a mixture of diastereomers.

   The *anti/syn* ratio (97:3) and the respective ee (*anti* 87%, *syn* 58%) were determined by chiral HPLC.

   ^{1}H NMR (270 MHz, CDCl$_3$) for *anti* isomer δ 0.88 (t, *J* 6.7 Hz, 3H), 1.26 (br, 26H), 1.60 (m, 2H), 2.55 (d, *J* 6.7 Hz, 1H), 2.63 (br, 1H), 3.82–3.98 (m,

2H), 4.02–4.18 (m, 2H), 7.10 (brd, 1H), 7.42–7.56 (m, 3H), 7.82 (d, J 6.7 Hz, 2H).

11.6   ENANTIOSELECTIVE REDUCTION OF KETONES USING N-ARYLSULFONYL OXAZABOROLIDINES

Mukund P. Sibi, Pingrong Liu, and Gregory R. Cook

Center for Main Group Chemistry, Department of Chemistry, North Dakota State University, Fargo ND, 58105–5516, Sibi@plains.nodak.edu, grcook@plains.nodak.edu

11.6.1   SYNTHESIS OF N-(2-PYRIDINESULFONYL)-1-AMINO-2-INDANOL

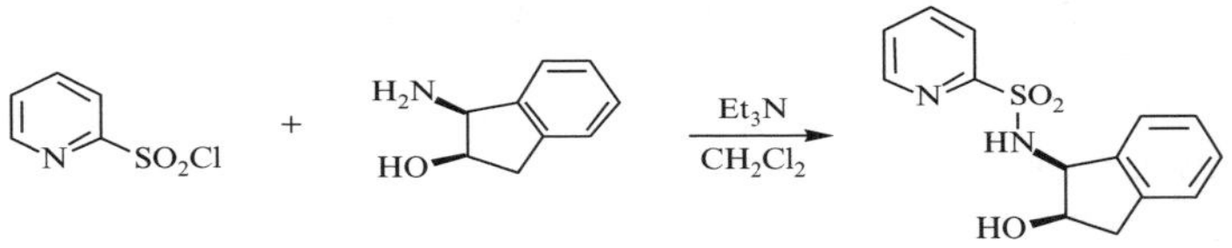

Materials and equipment

- Methylene chloride, 160 mL
- Triethylamine, 6.7 mL
- 2-Chlorosulfonyl pyridine, 7.1 g[21]
- (1S,2R) (Z)-Amino indanol, 5.97 g.
  Both enantiomers of (Z)-1-amino-2-indanol are available commercially.

- 250 mL Three-necked round-bottomed flask with a magnetic stirrer bar
- Magnetic stirrer
- Ice-bath

Procedure

1. The amino indanol was placed in a 250 mL three-necked round-bottomed flask equipped with a magnetic stirrer bar under nitrogen. Dry methylene chloride (110 mL) and triethylamine (6.7 mL) were then added. The reaction mixture was allowed to cool to 0 °C before adding a solution of 2-chlorosulfonyl pyridine (7.1 g in 50 mL CH$_2$Cl$_2$) over 20 minutes. The mixture was stirred at this temperature for 1 hour.
2. Water (60 mL) was added. The organic layer was separated. The aqueous layer was extracted with methylene chloride (4 × 100 mL). The organic layers

were combined and washed with brine, dried over magnesium sulfate and concentrated to give a white solid. The solid was purified by crystallization using ethyl acetate to give the product as white crystals (10 g, 87%).

^{1}H NMR (400 MHz, CDCl$_3$): δ 8.66 (d, J 4.8 Hz, 1H), 8.10 (d, J 8 Hz, 1H), 8.00 (m, 1H), 7.57 (m, 1H), 7.41 (dd, J 5.2, 7.6 Hz, 1H), 7.3–7.2 (m, 4H), 5.69 (d, J 8 Hz, 1H), 4.95 (dd, J 4.8, 9.7 Hz, 1H), 4.27 (m, 1H), 3.06 (dd, J 5.6, 16.7 Hz, 1H), 2.94 (d, J 16.6 Hz, 1H).

^{13}C NMR (100 MHz, CDCl$_3$): δ 158.8, 149.2, 140.0, 139.4, 139.1, 128.7, 127.3, 127.3, 125.3, 124.9, 122.5, 72.1, 62.3, 38.8.

Rotation was recorded on a JASCO-DIP-370 instrument: $[\alpha]_D^{25}$ − 37.0 (c 1.0, CHCl$_3$).

Analysis calculated for C$_{14}$H$_{14}$N$_2$O$_3$S: C, 57.92, H, 4.86, N, 9.65, Found: C, 57.67, H, 4.57, N, 9.65.

The quality of the ligand can be determined by performing an asymmetric reduction reaction on prochiral ketones according to the following procedure.

11.6.2 ASYMMETRIC REDUCTION OF A PROCHIRAL KETONE (CHLOROACETOPHENONE)

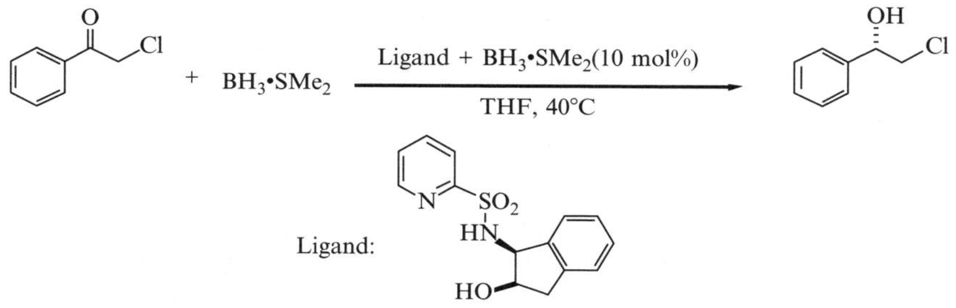

Materials and equipment

- Tetrahydrofuran, 71 mL
- 2-Chloroacetophenone, 1.02 g
- Borane-methyl sulfide complex (2 M in THF), 4.62 mL
- Ligand N-(2-Pyridinesulfonyl)-1-amino-2-indanol, 191.4 mg

- 100 mL Round-bottomed flask with a magnetic stirrer bar
- Magnetic stirrer hot plate
- Oil-bath
- Thermometer
- Syringe pump

Procedure

1. The ligand (191.4 mg) was placed in a 100 mL round-bottomed flask equipped with a magnetic stirrer bar in an oil-bath at 40 °C, under nitrogen. Dry tetrahydofuran (66 mL) was then added. After the solution turned clear, borane–methyl sulfide complex (1.32 mL) was added dropwise. The mixture was stirred at this temperature for 2.5 hours.

2. Borane–methyl sulfide complex (3.3 mL) was added to the reaction mixture. After stirring for an additional 1.5 hour at 40 °C, a solution of 2-chloroacetophenone (1.02 g in 5 mL of THF) was added over 2 hours using a syringe pump. The reaction was monitored by TLC and after completion (1.5 hour), it was cooled to 0 °C and quenched carefully with methanol. Solvent was removed on a rotary evaporator. 1 M HCl (15 mL) was added followed by extraction with dichloromethane (3 × 50 mL). The organic layers were combined and washed with brine, dried over magnesium sulfate, and concentrated to give a liquid.

3. The crude reaction mixture was purified by flash column chromatography (10 % ethyl acetate in hexane) to give the product as a colourless liquid (0.9 g, 90 % yield).
   - The ee (86–89 %) was determined by HPLC (Chiralcel OD column, flow rate 1 mL/min, eluent i-propanol–n-hexane 2:98), S-enantiomer: R_t 22.2 min, R-enantiomer: R_t 26.1 min.
   - ^{1}H NMR (400 MHz, CDCl$_3$): δ 7.4–7.3 (m, 5H), 4.89 (dd, J 13.2, 5.4 Hz, 1H), 3.8–3.6 (m, 2H), 2.62 (broad, 1H).
   - ^{13}C NMR (100 MHz, CDCl$_3$): δ 140.0, 128.8, 128.6, 126.2, 74.2, 51.0.

Conclusions

Oxazaborolidine-mediated reduction of ketones is very popular for the synthesis of enantiomerically pure secondary alcohols[22]. The present work illustrates an example of delivery of the hydride by borane coordinated to a remote Lewis basic site. The procedure is easy to reproduce. Slow addition of the ketone helps increase the enantioselectivity. The methodology is general and a variety of ketones can be reduced in high chemical yield and good enantioselectivity. The following table presents results from the reduction of a variety of ketones using the chiral ligand derived from amino indanol[23].

1-3   4   5

Table 11.8 Reduction of ketones using N-(2-pyridinesulfonyl)-
1-amino-2-indanol as a ligand.

Compound	X	Yield (%)	% ee (config.)
1	H	85	80 (R)
2	Br	91	87 (R)
3	OMe	90	77 (R)
4	–	90	87 (R)
5	–	83	71 (R)

11.7 REDUCTION OF KETONES USING AMINO ACID ANIONS AS CATALYST AND HYDROSILANE AS OXIDANT

MICHAEL A. BROOK

Department of Chemistry, McMaster University 1280 Main St. W. Hamilton, Ontario, Canada, L8S 4M1., Phone: (905) 525–9140 ext. 23483, Fax: (905) 522–2509, WWW: http://www.chemistry.mcmaster.ca/faculty/brook/brook.html

One of the fundamental operations in organic synthesis remains the stereoselective reduction of carbonyl groups[24]. In a process related to that reported by Hosomi *et al.*[25], using hydrosilanes as the stoichiometric oxidant and amino acid anions as the catalytic source of chirality, a variety of ketones were reduced in good to excellent yield and with good stereoselectivity[26]. This process reduces the amount of chiral catalyst needed and utilizes catalysts from the chiral pool that can be used directly in their commercially available form.

Materials and equipment

- L-Histidine, 50 mg, 0.3 mmol
- Dry tetrahydrofuran, 30 mL
- Distilled tetramethylethylene diamine, 1.0 mL, 6 mmol
- *n*-Butyllithium, 2 M solution in hexane, 0.32 mL, 0.6 mmol
- Trimethoxysilane (or triethoxysilane) 0.38 mL, 3 mmol
- Acetophenone 0.35 mL, 3 mmol
- Sodium hydrogen carbonate, 1 M solution, 20 mL
- Pentane (360 mL), diethyl ether (240 mL)
- Silica gel 60 (1 × 15 cm)
- Sand

- Two 100 mL one-neck round-bottomed flask
- Magnetic stirrer and stirrer bar
- Separatory funnel, 250 mL
- Rotary evaporator

Procedure

1. The 100 mL round-bottomed flask, equipped with a magnetic stirrer bar, was dried in an oven at 120 °C overnight. The flask was removed, sealed, cooled and flushed with nitrogen.
2. L-Histidine (50 mg) was placed in the flask. The flask was again flushed with nitrogen. Tetrahydrofuran (30 mL) was added and the mixture was stirred.
   L-Histidine is sparingly soluble in tetrahydrofuran.
3. To this stirring mixture at ambient temperature was added n-butyllithium (0.32 mL of a 2 M solution in hexane) dropwise. The resulting solution was stirred at ambient temperature for 30 minutes.
4. The clear mixture was cooled to 0 °C, and freshly distilled tetramethylene diamine (1 mL) was added. The system was stirred for 10 minutes after the addition.
   Tetramethylethylene diamine is hygroscopic.
5. Trimethoxysilane (0.38 mL) was added and the solution allowed to stir for an additional 10 minutes.

   Triethoxysilane and especially trimethoxysilane are rather toxic compounds (they may cause blindness if allowed to get into contact with eyes) and therefore care must be taken in their handing. Both need to be manipulated very carefully with suitable gloves, eyes face protection, in a well ventilated fume-hood. However, both can be handled without problems via syringe techniques.

   Although both triethoxysilane and trimethoxysilane are useful in these reactions, the latter reacts much more rapidly and, therefore is more convenient than the former.
6. Acetophenone (0.35 mL) was added and the resulting system was allowed to stir overnight at 0 °C.
7. The reaction was removed from the cooling bath and quenched with the addition of sodium hydrogen carbonate (20 mL), with vigorous stirring that was continued for 30 minutes at room temperature.
   Care must be taken in controlling the quenching time of the reaction. It was found that longer quenching times resulted in crude reaction mixtures that were difficult to effectively separate (lower product yields were obtained).
8. The biphasic system was transferred to a separatory funnel (250 mL) and extracted with ether (3 × 40 mL). The organic fractions were combined. The solvent was removed using a rotary evaporator, to produce a yellow oil and a white solid (polymerized trimethoxysilane).
9. The crude material was purified using flash silica gel chromatography eluting with pentane/ether (3:1). This provided 0.31 g (85%) phenethanol.
   ^{1}H NMR and/or ^{19}F NMR analysis of the Mosher ester of the resulting alcohol was used to determine the ee (25–30%).

General experimental procedure for preparation of Mosher esters[27]

For (*S*)-1-phenylethanol: (*S*)-1-phenylethanol (2 mg, 0.02 mmol) and MTPA-Cl (+) (4 mL, 0.02 mmol) were mixed with carbon tetrachloride (3 drops) and dry pyridine (3 drops). The reaction mixture was allowed to stand in a stoppered flask for 12 hours at ambient temperature. Water (1 mL) was added and the reaction mixture transferred to a separatory funnel and extracted with ether (20 mL). The ether solution, after washing successively with HCl (1 M, 20 mL), and saturated sodium carbonate solution (20 mL), and water (20 mL), was dried over sodium sulfate, filtered and solvents were removed *in vacuo*. The residue was dissolved in deuteriated chloroform for NMR analysis. The relative integration of the hydrogen on the carbon bearing the hydroxyl group was used to calculate the ee.

^{1}H NMR (CDCl$_3$, 200 MHz): δ 1.56 (d, 3H, J = 6.5 Hz, PhCH(OH)*CH*$_3$), 2.76 (bs, 1H, PhCH(*OH*)CH$_3$), 4.94 (q, 1H, J = 6.5 Hz, Ph*CH*(OH)CH$_3$), 7.32–7.45 (m, 5H$_{arom}$).

^{13}C NMR (CDCl$_3$, 200 MHz): δ 24.97, 69.99, 125.24, 127.14, 128.24, 145.75.

FTIR (neat, KBr disc) ν (cm^{-1}) 3364, 3065, 3031, 2974, 2929, 1728, 1603, 1494, 1452, 1371, 1287, 1204, 1077, 1030, 1011, 900, 762, 700, 607, 541.

Conclusion

The stereoselective reduction may be applied to a variety of ketones. Some examples of reductions, as a function both of ketone substrate and amino acid catalyst are provided in Table 11.9. The full scope of this procedure[26–28] has

Table 11.9   Reduction of Ketones Using HSi(OEt)$_3$ and amino acid anions.

R^1	R^2	Amino acid anion (mol%)	Yield %	ee%
H	Me	Li$_2$-His (10)	85	26 (*S*)
H	Me	Li-His (10)	75	26 (*S*)
CF$_3$	Me	Li$_2$-His (10)	86	30 (*S*)
Me	Me	Li$_2$-His (10)	80	40 (*S*)
Me	Ph	Li$_2$-His (10)	82	5 (*S*)
CF$_3$	Ph	Li$_2$-His (10)	95	30 (*S*)
H	Me	Li-Phe (100)	70	25 (*S*)

not been completely mapped out and, in particular, the use of other amino acids such as proline, which are known to be particularly useful chiral catalysts[29], must be examined.

REFERENCES

1. Hirao, A., Itsuno, S., Nakahama, S., Yamasaki, N. *J. Chem. Soc., Chem. Commun.*, 1981, 315.
2. Brunel, J.M., Pardigon, O., Faure, B., Buono, G. *J. Chem. Soc., Chem. Commun.*, 1992, 287.
3. Chiodi, O., Fotiadu, F., Sylvestre, M., Buono, G. *Tetrahedron Lett.,* 1996, **37**, 39.
4. Gamble, M.P., Studley, J.R., Wills, M. *Tetrahedron Lett.,* 1996, **37**, 2853.
5. Bolm, C. *Angew. Chem. Int. Ed. English*, 1993, **32**, 232.
6. Corey, E.J., Bakshi, R.K., Shibata, S. *J. Am. Chem. Soc.,* 1987, **109**, 5551.
7. Corey, J., Bakshi, R.K., Shibata, S., Chen, H.-P., Singh, V. K. *J. Am. Chem. Soc.,* 1987, **109**, 7925.
8. Corey, E.J., Chen, C.-P., Reichard, G. A. *Tetrahedron Lett.,* 1989, **30**, 5547.
9. Corey, E.J., Link, J.O. *Tetrahedron Lett.,* 1989, **30**, 6275.
10. Corey, E.J., Helal, C.J. *Tetrahedron Lett.,* 1996, **37**, 5675.
11. Wallbaum, S., Martens, J. *Tetrahedron: Asymmetry*, 1992, **3**, 1475.
12. Salunkhe, A.M., Burkhardt, E.R. *Tetrahedron Lett.,* 1997, **38**, 1523.
13. Buono, G., Chiodi, O., Wills, M. *Synlett* 1999, 377.
14. Bolm, C., Felder, M. *Tetrahedron Lett.,* 1993, **34**, 6041.
15. (a) Hong, Y., Gao, Y., Nie, X., Zepp, C.M. *Tetrahedron Lett.,* 1994, **35**, 6631. (b) Ghosh, A.K., Fidanze, S., Senanayake, C.H., *Synthesis*, 1998, 937. (c) Senanayake, C.H. *Aldrichimica Acta*, 1998, **31**, 1 and references cited therein.
16. (a) Hett, R., Fang, Q.K., Gao, Y., Wald, S.A., Senanayake, C.H., *Org. Proc Res and Dev.*, 1998, **2**, 96. (b) Hett, R., Wald, S.A., Senanayake, C.H. *Tetrahedron Lett.,* 1998, **39**, 1705.
17. (a) Jones, T.K., Mohan, J.J., Xavier, L., Blacklock, T.J., Mathre, D.J., Sohar, P., Tunner Jones, E.T., Reamer, R.A., Grabowski, E.J. *J. Org. Chem.,* 1991, **56**, 763; (b) Salunkhe, A.M., Burkhardt, E.R. *Tetrahedron Lett.,* 1997, **38**, 1523.
18. Senanayake, C.H., DiMichele, L.M., Liu, J., Fredenburgh, L.E., Ryan, K.M., Roberts, F.E., Larsen, R.D., Verhoeven, T.R., Reider, P.J. *Tetrahedron Lett.,* 1995, **36**, 7615.
19. Inspection of molecular model of oxazaborolidine **2a** and **5a** indicated that the distance of hydrogen (C_{ortho} − H) to boron (N-BH$_3$) of **5a** is closer than **2a**.

2a   5a

20. The B-methyl catalysts were prepared by reacting the aminoalcohol with trimethylboroxine, followed by an azeotropic distillation with toluene.
21. For the synthesis of 2-chlorosulfonyl pyridine see: Diltz, S., Aguirre, G., Ortega, F., Walsh, P.J. *Tetrahedron: Asymmetry*, 1997, **8**, 3559.
22. For a comprehensive and excellent review see: Corey, E.J., Helal, C. *Angew. Chem. Int. Ed. English*, 1998, **37**, 1986.
23. Sibi, M.P., Cook, G.R., Liu, P. *Tetrahedron Lett.,* 1999, **40**, 2477.
24. Smith, M.B. *Organic Synthesis*, McGraw Hill: New York, 1994, Chap. 4,5, pp. 343–505.
25. (a) Kohra, S., Hayashida, H., Tominaga, Y., Hosomi, A. *Tetrahedron Lett.*, 1988, *29*, 89. (b) Hojo, M., Fuji, A., Murakami, C., Aihara, M., Hosomi, A. *Tetrahedron Lett.*, 1995, **36**, 571.
26. Laronde, F.J., Brook, M.A. *Tetrahedron Lett.*, 1999, **40**, 3507.
27. Yamaguchi, S. *Asymmetric Synthesis*, Morrison, J.D. Ed., Academic: New York, 1983, Vol. 1, p 128; Niwa, H., Ogawa, T., Okamato, O., Yamada, K. *Tetrahedron*, 1992, **48**, 10531; Latpov, S.K., Seco, J.M., Quinoa, E., Riguera, R. *J. Org. Chem.*, 1995, **50**, 504; *idem. ibid.* 1996, **61**, 8569; *idem. ibid. J. Am. Chem. Soc.*, 1998, **120**, 877.
28. LaRonde, F.J., Brook, M.A. *Stereoselective Reduction of Ketones Using Extracoordinate Silicon: C_2-Symmetric Ligands, Inorganica Chim. Acta.*
29. Proline in Michael additions: Tomioka, K., Koya, K. *Asymmetric Synthesis*, Morrison, J.D., Ed., Academic: New York, 1983, pp. 219–21. Proline in aldol reactions: Hajos, Z.G., Parrish, D.R. *Org. Synthesis.* 1985, **63**, 26. Sauer, G., Eder, U., Haffer, G., Neef, G. *Angew. Chem., Int. Ed. English*, 1975, **14**, 417.

12 Asymmetric Hydrogenation of Carbon–Carbon Double Bonds Using Organometallic Catalysts

CONTENTS

12.1   INTRODUCTION

Acrylic acid derivatives were chosen as substrates in the early studies on asymmetric hydrogenation of olefins. The additional coordinating functionality such as an amido, carboxyl, amidomethyl, carbalkoxymethyl or hydrocarbonylmethyl group is a prerequisite for getting higher enantioselectivities. Concerning the synthesis of enantiomerically pure α-amino acids, chiral rhodium–diphosphines[1–3] or ruthenium-[4] catalysts were found to give good results for the enantioselective hydrogenation of α-(acetamido)acrylates α-(enamides) (Figure 12.1). In this section catalytic hydrogenation methods using ligands based on chiral templates, such as [1,2-bis(phospholano)benzene] (DuPHOS), [1,2-(bisphospholano)ethane] (BPE) and 3,6-bis[bis(4-fluorophenyl)-phosphinoxy]-bicyclo[3.2.0]-heptane (B [3.2.0]DPO) will be described.

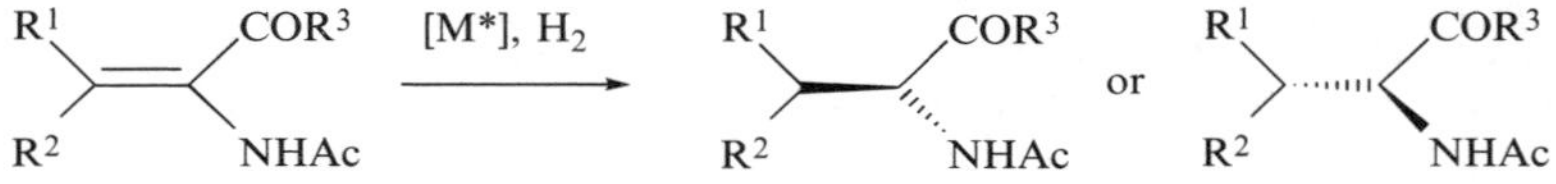

Figure 12.1   Asymmetric hydrogenation of α-enamides.

One of the possible catalytic cycles (i.e. for olefin hydrogenation) is described in Figure 12.2. The molecular hydrogen is first complexed to the metal. Then the olefin is complexed and inserted into the M–H bond. The alkane is liberated by elimination and the catalyst regenerated.

Extreme caution must be taken whenever hydrogen gas and active catalyst are used. Never allow naked flames in the vicinity when hydrogen is being used. Avoid the formation of air–hydrogen mixtures. Any electrical apparatus in the vicinity must be spark-proof. It is far better for the apparatus to be kept in a separate room specially designed for hydrogenation[5].

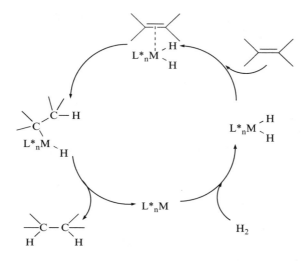

Figure 12.2 Mechanism of olefin hydrogenation by transition metal complexes.

12.2 HYDROGENATION OF DIMETHYL ITACONATE USING [Rh((*S,S*)-Me-BPE)][6]

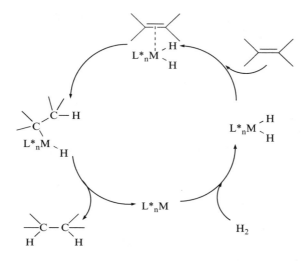

$\left(\begin{smallmatrix} P \\ P \end{smallmatrix} \right.$ = (*S,S*)-Me-BPE

Materials and equipment

- Dimethyl itaconate, 177 μL, 200 mg, 1.26 mmol
- (*S,S*)-1,2-bis(1′,4′-Dimethylphospholano)ethane(cyclooctadiene)rhodium(I): [(COD)Rh(*S,S*)-Me-BPE], 0.7 mg, 1.26 μmol, 0.1 mol%*

---

* *The catalyst [(COD)Rh((S,S)-Me-BPE)] was kindly provided by Dr M. Burk (Chirotech Technology Limited, Cambridge UK).*

- Anhydrous methanol degassed for 20 minutes with nitrogen, 5 mL
- Ethyl acetate, diethyl ether
- Silica gel 60 (0.063–0.04 mm)

- 10 mL Glass liner adapted to the high pressure reactor
- Magnetic stirrer hot plate with a thermostatically controlled oil-bath and thermometer
- 25 mL High pressure reactor
  The reaction can be performed at atmospheric pressure using a low-pressure hydrogenation apparatus fitted with a gas burette system.

Procedure

1. A 10 mL glass liner equipped with a magnetic stirrer bar was dried in an oven at 120 °C overnight, cooled in a desiccator under vacuum and then flushed with nitrogen.
2. The liner was filled under nitrogen with dimethyl itaconate (177 μL, 200 mg) and the catalyst [(COD)Rh(S,S)-Me-BPE] (0.7 mg) and then placed in a 25 mL high pressure reactor.
3. The reactor was flushed six times with hydrogen (the bomb was pressurized at 200 psi, then the gas inlet was closed before the hydrogen was slowly vented off). Degassed anhydrous methanol (5 mL) was added and the reactor was pressurized to an initial pressure of 50 psi H_2. The reaction was allowed to stir at 20 °C until no further hydrogen uptake was observed (2 hours).
4. The reaction was followed by chiral GC (SE 30, 100 °C isotherm, nitrogen mobile phase). R_t (dimethyl itaconate): 6.8 min; R_t (dimethyl methylsuccinate): 5.7 min.
5. The reaction was then concentrated and the residue was passed through a short column of silica gel eluting with ethyl acetate–diethyl ether (1:1) to remove the catalyst. The (S)-dimethyl methylsuccinate does not need any further purification (190 mg, 95 %).
   - The ee (95 %) was determined by chiral GC (Lipodex® E, 25 m, 0.25 mm ID, temperatures: column 75 °C isotherm, injector 250 °C, detector 250 °C, mobile phase helium, sample dissolve in methanol) R_t (R)-enantiomer: 39.3 min R_t (S)-enantiomer: 41.2 min.
   - ^{1}H NMR(200 MHz, CDCl$_3$): δ3.70 (s, 3H, CO$_2$CH_3); 3.69 (s, 3H, CO$_2$CH_3); 2.92 (m, 1H, CH); 2.76 (dd, J 16.5 Hz, J 8.2 Hz, 1H, CH_aH$_b$); 2.42 (dd, J 16.5 Hz, J 6.0 Hz, 1H, CH$_a$$H_b$); 1.23 (d, J 7.1 Hz, 3H, CH_3).
   - IR (CHCl$_3$, cm^{-1}): 3031, 2957 (C–H aliphatic), 1727 (C=O), 1463, 1437, 1353, 1280 (C–O), 1167, 1059, 1007.
   - Mass: calculated for C$_7$H$_{13}$O$_4$: m/z 161.08138; found [MH]$^+$ 161.08154.

12.3 HYDROGENATION OF AN α-AMIDOACRYLATE USING [Rh ((R,R)-Me-DuPHOS)][2]

Materials and equipment

- α-Acetamido cinnamic acid, 259 mg, 1.26 mmol
- (−)-(R,R)-1,2-bis(1',4'-dimethylphospholane)benzene(cyclooctadiene)rhodium (I): [(COD)Rh((R,R)Me-DuPHOS)], 0.8 mg, 1.26 μmol, 0.1 mol%*
    This catalyst is commercially available from Strem or Chirotech.
- Anhydrous methanol degassed for 20 minutes with nitrogen, 5 mL
- Ethanol, petroleum ether

- 10 mL Glass liner adapted to the 25 mL high pressure reactor with a magnetic stirrer bar
- 25 mL High pressure reactor
    The reaction can be performed at atmospheric pressure using a low-pressure hydrogenation apparatus fitted with a gas burette system.

Procedure

1. A 10 mL glass liner equipped with a magnetic stirrer bar was dried in an oven at 120 °C overnight, cooled in a desiccator under vacuum and then flushed with nitrogen.
2. The liner was filled under nitrogen with acetamido cinnamic acid (259 mg), anhydrous methanol (5 mL) and catalyst [(COD)Rh(R,R)-Me-DuPHOS] (0.8 mg). The liner was placed in a 25 mL high pressure reactor.
3. The reactor was flushed six times with hydrogen (the bomb was pressurized at 200 psi, then the gas inlet was closed before the hydrogen was slowly vented off) and then pressurized to an initial pressure of 90 psi H$_2$. The

---

* The catalyst [(COD)Rh((R,R)Me-DuPHOS)] was kindly provided by Dr M. Burk (Chirotech Technology Limited, Cambridge UK)

reaction was allowed to stir at 20 °C until no further hydrogen uptake was observed (3 hours).

4. The reaction was followed by chiral GC (SE 30, 220 °C, nitrogen mobile phase). R_t (α-acetamido cinnamic acid): 3.70 min; R_t (N-acetyl-L-phenyl-alanine): 5.4 min.

5. The reaction was concentrated to give a yellow oil (300 mg) which was crystallized with ethanol and petroleum ether to give slightly yellow crystals (235 mg, 90 %).

The ee (>98 %) was determined by chiral HPLC (Chiralpak® AD, Hexane-IPA-TFA, 89 %–10 %–1 %, sample dissolved in IPA) R_t (R)-enantiomer: 11.9 min, R_t (S)-enantiomer: 14.3 min.

^{1}H NMR(200 MHz, DMSO): δ8.25 (d, J 8.2 Hz, 1H, NH); 7.26 (m, 5H, Ph); 4.42 (m, 1H, CH); 3.07 (dd, J 13.7 Hz, J 4.9 Hz, 1H, CH$_a$H$_b$); 2.85 (dd, J 13.7 Hz, J 9.9 Hz, 1H, CH$_a$H$_b$); 1.75 (s, 3H, CO-CH$_3$).

Mass: calculated for $C_{11}H_{13}O_4N$: m/z 207.08954; found [MH]$^+$ 207.08975.

Conclusion

The procedures using [(COD) Rh (S, S)-Me-BPE] and [(COD) Rh (R,R)-Me-DuPHOS] are very similar; they need a hydrogenation bomb and are conducted under an inert atmosphere, as the catalysts are sensitive to oxygen. They give good results (yield and enantiomeric excess) and hydrogenated products do not need lengthy purification, since no secondary products were detected. The reactions can be carried out under atmospheric pressure giving approximately the same results but need a longer time to be complete. The reaction were stopped when no more hydrogen was consumed; they were generally performed overnight (14 hours). Table 12.1 gives some examples of β, β-disubstituted enamides that can be hydrogenated by those catalysts in similar conditions.

12.4 HYDROGENATION OF AN α-AMIDOACRYLATE USING [Rh(B[3.2.0]DPO)] COMPLEXES

**12.4.1 PREPARATION OF (COD)₂Rh⁺BF₄⁻*

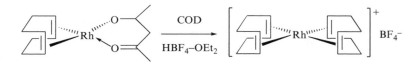

Materials and equipment

- [(COD)Rh(acac)], 3.1 g, 10 mmol
- Anhydrous tetrahydrofuran, 60 mL

---

* Dr. C. Dousson and Dr N. Derrien (University of Liverpool, UK) provided the procedures for the catalyst synthesis and the hydrogenation with Rh(B[3.2.0]DPO)[7].

- Cycloocta-1,5-diene, 1.30 g, 12 mmol, 1.2 eq
- Tetrafluoroboric acid–diethyl ether complex (HBF$_4$.OEt$_2$) in diethyl-ether, 54%, 3.00 g, 2.52 mL, 10 mmol, 1 eq, diluted with tetrahydrofuran, 5 mL
- Dry diethyl ether

- 100 mL Schlenk tube with a magnetic stirrer bar.
- Condenser.
- Magnetic stirrer hot plate with a thermostatically controlled oil-bath and thermometer
- Sinter funnel with an inert gas inlet

Table 12.1 Hydrogenation of β, β-disubstituted enamides by [(COD)Rh (S,S)-Me-BPE] and [(COD)Rh (R,R)-Me-DuPHOS] catalysts[2] (results according to the literature).

	[Rh(S,S)-Me-DuPHOS] ee % (configuration)	[Rh(R,R)-Me-BPE] ee % (configuration)
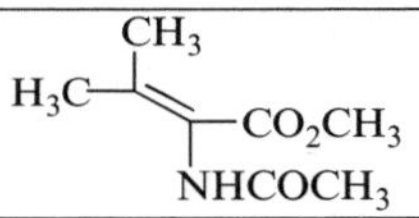 H$_3$C–C(CH$_3$)=C(CO$_2$CH$_3$)(NHCOCH$_3$)	96.0 (S)	98.2 (R)
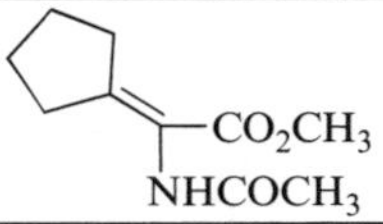 cyclopentylidene–C(CO$_2$CH$_3$)(NHCOCH$_3$)	96.8 (S)	97.2 (R)
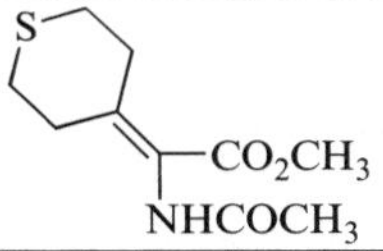 thiopyranylidene–C(CO$_2$CH$_3$)(NHCOCH$_3$)	95.0 (S)	98.4 (R)
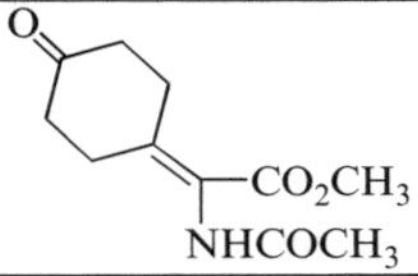 oxo-cyclohexylidene–C(CO$_2$CH$_3$)(NHCOCH$_3$)	93.7 (S)	98.0 (R)
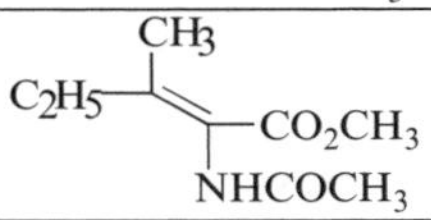 C$_2$H$_5$–C(CH$_3$)=C(CO$_2$CH$_3$)(NHCOCH$_3$)	–	98.2 (2R,3S)
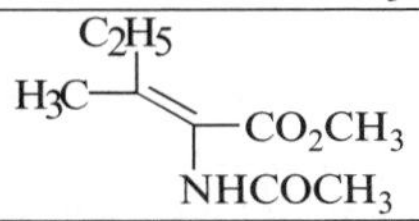 H$_3$C–C(C$_2$H$_5$)=C(CO$_2$CH$_3$)(NHCOCH$_3$)	–	98.3 (2R,3R)
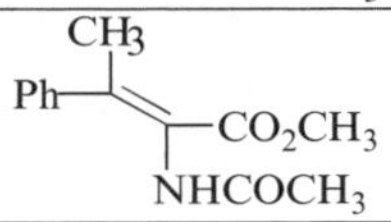 Ph–C(CH$_3$)=C(CO$_2$CH$_3$)(NHCOCH$_3$)	99.4 (2S,3R)	80.1 (2R,3R)

Procedure

1. A 100 mL Schlenk flask equipped with a magnetic stirrer bar and a condenser was dried at 150 °C overnight, cooled under vacuum and then flushed with nitrogen.
2. The Schlenk tube was filled with [(COD)Rh(acac)] (3.1 g) and cycloocta-1,5-diene (1.30 g) which were dissolved in 15 mL of dry tetrahydrofuran. To this orange mixture, the solution of HBF$_4$.OEt$_2$ in tetrahydrofuran (7.52 mL) was added. A brown precipitate appeared, giving a viscous solution, which was diluted with 40 mL additional tetrahydrofuran to allow the reaction to stir efficiently.
3. The orange solution was heated (80 °C) to reflux under nitrogen for 30 minutes.
4. The brown solution was cooled to room temperature. The brown powder was filtered under nitrogen using a sinter funnel with an inert gas inlet, and then washed with dry diethyl ether (3 × 5 mL).
5. The [(COD)$_2$Rh$^+$BF$_4^-$] complex obtained was used as the catalyst precursor for hydrogenation without further purification.

12.4.2 PREPARATION OF THE BISPHOSPHINITE LIGAND

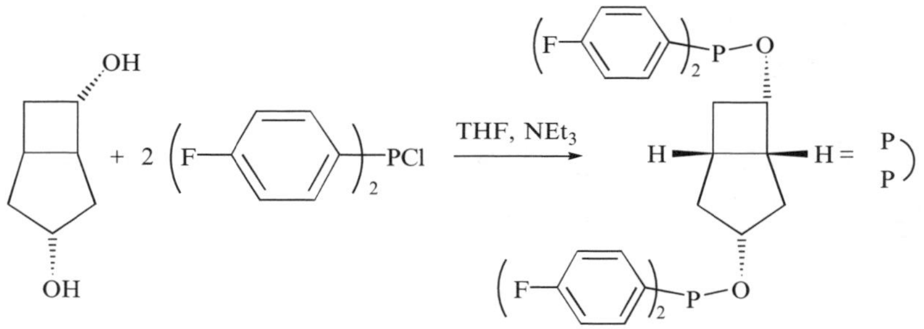

Materials and equipment

- (−)-(1R,3R, 5R, 6S)-Bicyclo[3.2.0]heptan-3,6-diol, 500 mg, 3.9 mmol
- Anhydrous tetrahydrofuran, 20 mL
- Triethylamine, 0.87 g, 8.6 mmol, 2.2 eq
- Bis(4-fluorophenyl)chlorophosphine, 2.2 g, 8.6 mmol, 2.2 eq

> *Chlorophosphines need to be manipulated carefully with gloves and eye protection. They can cause burns, irritation to eyes and irritation to the respiratory system.*

Different chlorophosphines can be synthesised, or are available from Strem or Digital Chemicals.

- Alumina, activated overnight at 150 °C

- Two 100 mL Schlenk tubes with magnetic stirrer bars
- Ice-bath
- Sinter funnel with an inert gas inlet

Procedure

1. Two 100 mL Schlenk flasks, each equipped with a magnetic stirrer bar, were dried at 150 °C overnight, cooled under vacuum and then flushed with nitrogen.
2. One of the Schlenk tubes was filled with $(-)$-$(1R, 3R, 5R, 6S)$-bicyclo [3.2.0] heptan-3,6-diol (500 mg) dissolved in dry tetrahydrofuran (20 mL) under nitrogen, then triethylamine (0.87 g) was added.
3. The solution was cooled to 0 °C in an ice-bath, and then the chlorophosphine (2.2 g) was added dropwise via a syringe over 5 minutes with stirring. A white precipitate of triethylammonium chloride appeared. When the addition was complete, the ice-bath was removed and the stirring was continued at ambient temperature for 15 hours.
4. A sinter funnel with nitrogen inlet connected to the second dry Schlenk tube was filled with a pad of activated alumina which was cooled under vacuum and then flushed with nitrogen. The precipitate was filtered off through the pad of alumina under nitrogen. The solvent was removed under vacuum from the second Schlenk tube.
5. The solvent was removed *in vacuo* to give the bisphosphinite ligand $(1R, 3R, 6S)$-3,6-bis[bis(4'-fluorophenyl) phosphinooxy] bicyclo [3.2.0] heptane as a white solid (1.99 g, 90%).

The ligands prepared by this method were sufficiently pure for use as an *in situ* catalyst preparation.

NMR ^{13}C (50 MHz, CDCl$_3$): δ 30.16 (s, C$_1$); 33.80 (d, $^3J_{PC}$ 5.4 Hz, C$_4$), 37.04 (d, $^3J_{PC}$ 6.5 Hz, C$_7$), 40.88 (d, $^3J_{PC}$ 4.8 Hz, C$_2$); 45.43 (d, $^3J_{PC}$ 4.8 Hz C$_5$); 69.98 (d, $^2J_{PC}$ 16.3 Hz, C$_6$); 84.78 (d, $^2J_{PC}$ 17.7 Hz, C$_3$); 115.18–116.88 (m, C$_{3'}$, C$_{5'}$, C$_{3''}$, C$_{5''}$); 131.89–133.87 (m, C$_2'$, C$_6'$, C$_2''$, C$_6''$); 136.99–137.83 (m, C$_{1'}$, C$_{1''}$); 163.54, 163.63, 163.71, 163.77 (4 × d, $^1J_{FC}$ 247 Hz, C$_{4'}$, C$_{4''}$).

NMR ^{31}P (162 MHz, CDCl$_3$): δ 103.30 (s), 105.03 (s).

Mass: calculated for C$_{31}$H$_{26}$F$_4$O$_2$P$_2$: m/z 568.13446; found [M]$^+$ 568.13466.

12.4.3　ASYMMETRIC REDUCTION OF α-ACETAMIDO CINNAMIC ACID

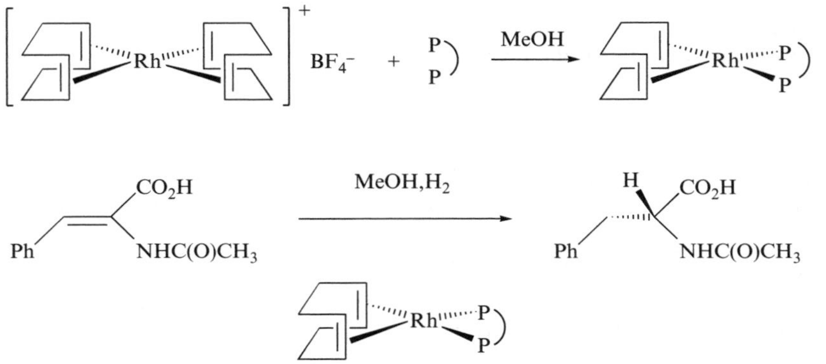

Materials and equipment

- (1R, 3R, 5R, 6S)-3, 6-Bis [bis (4′-fluorophenyl) phosphinoxy] bicyclo [3.2.0] heptane, 6.7 mg, 0.012 mmol, 1 mol%
- Anhydrous methanol degassed with nitrogen, bubbling for 1 hour, 30 mL $(COD)_2Rh^+BF_4^-$, 5.25 mg, 0.013 mmol, 1.1 mol%
  The catalyst is not stable in solution and cannot be stored for a long time.
- α-Acetamido cinnamic acid, 240 mg, 1.17 mmol

- 25 mL Schlenk tube with a magnetic stirrer bar
- Syringes
- High pressure reactor, 50 mL.
- Glass liner adapted to high pressure reactor with a magnetic stirrer bar

Procedure

1. A 25 mL Schlenk tube equipped with a magnetic stirrer bar was dried at 150 °C overnight, cooled under vacuum and then flushed with nitrogen.
2. The Schlenk tube was filled with bisphosphinite ligand, (1R, 3R, 5R, 6S)-3,6-bis [bis (4′-fluorophenyl) phosphinooxy] bicyclo[3.2.0]heptane (6.7 mg), degassed methanol (3 mL) and $(COD)_2Rh^+BF_4^-$ (5.25 mg). The reaction mixture was stirred at room temperature until all the material was dissolved (10–15 minutes) giving an orange solution.
3. A glass liner of a 50 mL hydrogenation bomb was charged with α-acetamido cinnamic acid (240 mg) and a magnetic stirrer bar. The bomb was then assembled, flushed five times with hydrogen (the bomb was pressurized at 200 psi, then the gas inlet was closed before the hydrogen was slowly vented off).

4. The solution of the catalyst (formed *in situ*) was added via a syringe (3 mL) through the solvent port equipped with a septum, and the mixture stirred.
5. The hydrogenation bomb was pressurized to 200 psi of hydrogen (14 atm). The reaction performed at room temperature was complete after 3 hours (followed by GC/MS). N-Acetyl-L-phenylalanine was obtained in quantitative yield.

The ee (91 %) was determined by chiral HPLC (Chiralpak® AD, Hexane–IPA–TFA, 89%–10%–1%, sample dissolved in IPA) R_t (R)-enantiomer: 11.9 min, R_t (S)-enantiomer: 14.3 min.

^{1}H NMR (200 MHz, DMSO): δ 8.22 (d, J 8.2 Hz, 1H, NH); 7.24 (m, 5H, Ph); 4.40 (m, 1H, CH); 3.02 (dd, J 13.8 Hz, J 5.0 Hz, 1H, CH$_a$$H_b$); 2.83 (dd, J 13.8 Hz, J 9.5 Hz, 1H, CH_aH$_b$); 1.78 (s, 3H, CO-CH_3).

Other ligands were synthesis by the same methods using different chlorophosphines. The reduction reaction of the α-acetamido cinnamic acid gave good results in term of enantiomeric excess and yield (all the reactions went to completion). The results are summarised in Table 12.2.

Conclusion

The rhodium–diphosphine catalysts are generally sensitive to oxygen, hence the reactions have to be carried out under strictly inert atmospheric conditions. A decrease in the yield or the enantiomeric excess can be due to a lack of sufficient precaution during the procedure or to the inactivation of the catalyst when exposed to oxygen. However, the reactions using rhodium complexes as catalysts give very good results which correlate well with the published material.

Table 12.2   Enantiomeric excess resulting from the reduction of α-acetamido cinnamic acid by rhodium (B[3.2.0]DPO) complexes.

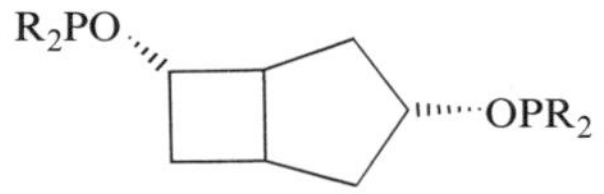

Ligand's substituent, R	ee %
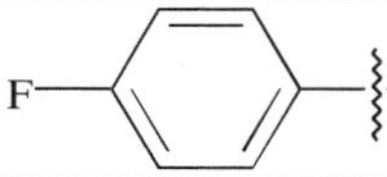	91
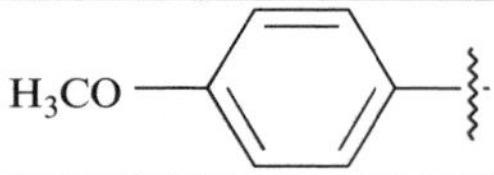	90.5
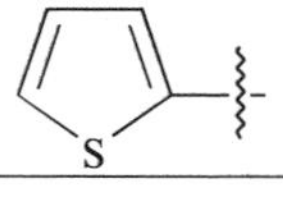	87.5

Note that, in contrast, the reactions using [(COD) Rh ((S, S)Me-BPE)] or [(COD)Rh((R, R) Me-DuPHOS)] complexes can be performed at atmospheric pressure of hydrogen which avoids the use of heavy-duty hydrogenation apparatus.

12.5 HYDROGENATION OF ENOL CARBONATES AND 4-METHYLENE-N-ACYLOXAZOLIDINONE USING [Rh((R)-BiNAP)] COMPLEXES

P.H. Dixneuf, C. Bruneau and P. Le Gendre

UMR6509, Organometalliques et Catalyse: Chimie et Electrochimie Moleculaire, Universite de Rennes 1, Laboratoire de Chimie de Coordination Organique, Campus de Beaulieu, Avenue du général Leclerc, 35042 Rennes Cedex, Tel: + 33 (0)2 99 28 62 80, Fax: + 33 (0)2 99 28 69 39, e-mail: pierre.dixneuf@univ-rennes 1. fr

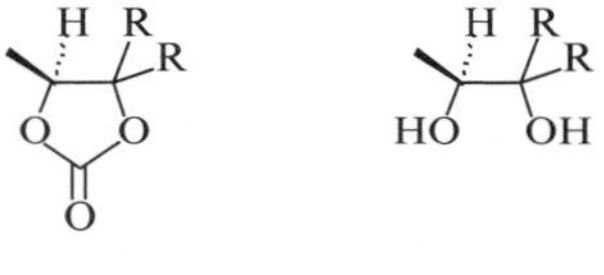

Figure 12.3 $R = Me, R - R = -(C_2H)_4^-, R - R = -(C_2H)_5^-$.

P.H. Dixneuf, C. Bruneau and P. Le Gendre[8] have reported a straightforward synthesis of optically active cyclic carbonates and 1,2-diols (Figure 12.3) based on the selective hydrogenation of the exocyclic double bond of α-methylene carbonates[8,9] followed by their hydrolysis. By using bis(trifluoroacetate) BiNAP-ruthenium[10] complexes as precatalyst, the asymmetric hydrogenation of α-methylene-1,3-dioxolan-2-ones was carried out in dichloromethane solution under 10 MPa hydrogen pressure. This procedure allowed access to cyclic carbonates with high yields (80–85%) and optical purities (89–95%). The treatment of these carbonates with potassium carbonate in anhydrous methanol for 2.5 hours led to the quantitative conversion of the carbonates into the corresponding diols.

12.5.1 SYNTHESIS OF (S)-4,4,5-TRIMETHYL-1,3-DIOXOLANE-2-ONE

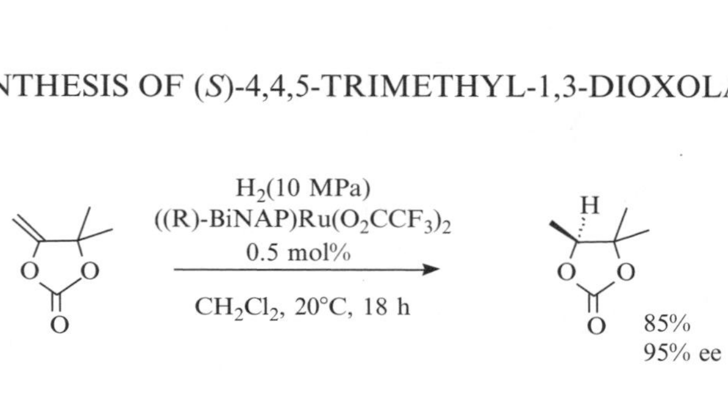

Materials and equipment

- 5-Methylene-1,3-dioxolane-2-one[9], 0.25 g, 1.95 mmol
- ((R)-BiNAP)Ru(O$_2$CCF$_3$)$_2$[10], 9 mg, 0.01 mmol
- Dry and degassed dichloromethane, 15 mL

- 125 mL Stainless steel autoclave with a mechanical stirrer
- 50 mL Round bottomed flask
- Rotary evaporator
- Kugelrohr apparatus

Procedure

1. The 125 mL stainless steel autoclave was flushed with nitrogen. The 5-methylene-1,3-dioxolane-2-one, the ruthenium catalyst and dichloromethane (10 mL) were placed in the autoclave under a nitrogen atmosphere.
2. The autoclave was sealed, flushed with hydrogen and pressurized with 10 MPa of hydrogen. The mixture was stirred for 18 hours at 20 °C under 10 MPa of hydrogen.
3. Once the autoclave was depressurized, the solution was poured into a 50 mL round bottomed flask and the autoclave rinsed with dichloromethane (5 mL). The solvent was removed by using a rotary evaporator.
4. The hydrogenated carbonate can be recovered free of ruthenium catalyst by sublimation under reduced pressure using a Kugelrohr apparatus (bp 70 °C, 1.5 mmHg).

   This procedure has been scaled up to provide 2 g of 4,4,5-trimethyl-1,3-dioxolane-2-one.

   The optical purity can be determinated by using GC with a chiral Lipodex capillary column (25 m × 0.25 mm).

12.5.2   SYNTHESIS OF (S)-2-METHYL-2,3-BUTANEDIOL

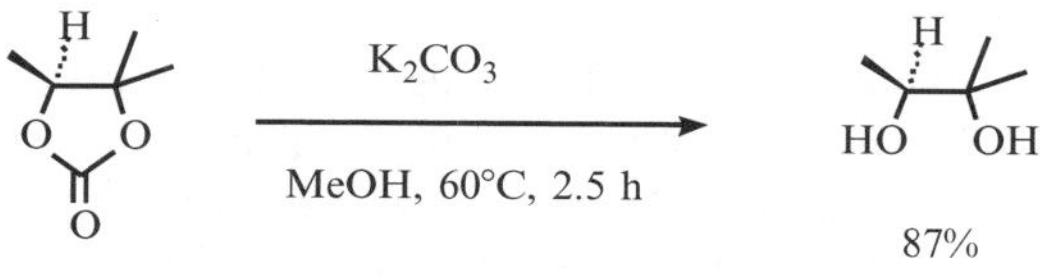

Materials and equipment

- 4,4,5-Trimethyl-1,3-dioxolane-2-one, 0.17 g, 1.34 mmol
- Potassium carbonate, 0.27 g, 2.0 mmol
- Dry methanol, 10 mL
- Diethyl ether, 10 mL
- Saturated solution of NH$_4$Cl, 5 mL
- Magnesium sulfate

- 50 mL Round bottomed flask with a magnetic stirrer bar
- Reflux condenser
- Magnetic stirrer plate with thermostatically controlled oil bath and thermometer
- Rotary evaporator
- Kugelrohr apparatus

Procedure

1. 4,4,5-Trimethyl-1,3-dioxolane-2-one (0.17 g), potassium carbonate (0.27 g) and methanol (10 mL) were placed in 50 mL round bottomed flask equipped with a magnetic stirrer bar and a reflux condenser. The mixture was then stirred at 60 °C for 2.5 hours.
2. The solvent was removed by using a rotary evaporator. The solution was dissolved in a saturated solution of NH_4Cl and extracted with diethyl ether. After the solution was dried with magnesium sulfate, the diethyl ether was removed by using a rotary evaporator.
3. It is noteworthy that this diol has been used as ligand in the molybdenum-mediated kinetic resolution of oxiranes[11].

12.5.3   PREPARATION OF OPTICALLY ACTIVE N-ACYLOXAZOLIDINONES

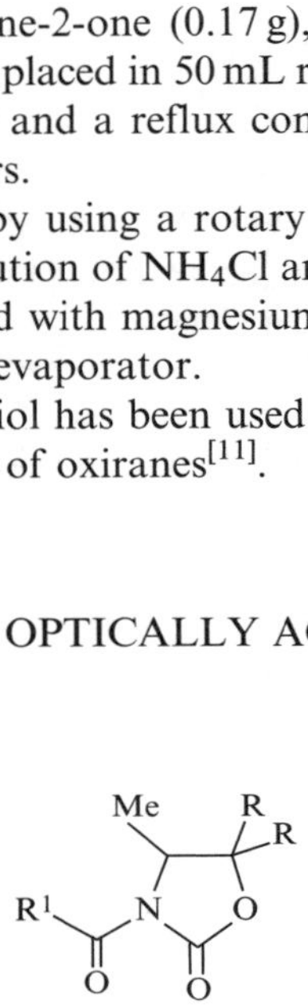

Figure 12.4 R^1 = Me, Et, Ph; R = Me, R-R= -(C_2H_5).

Whereas optically active acyloxazolidinones are usually prepared by acylation of oxazolidinone arising from optically active natural amino acids via multistep synthesis[12], Dixneuf's research group[13] recently described a novel route to both enantiomers of optically active N-acyloxazolidinones (Figure 12.4) via asymmetric hydrogenation of 4-methylene-N-acyloxazolidinones[13,14]. The enantioselective hydrogenation of the latter was performed under 10 MPa of hydrogen in MeOH at 50 °C for 18 hours in the presence of 1 mol% of ((R)-BiNAP)Ru$(O_2CCF_3)_2$,[10] as catalyst and led to optically active N-acyloxazolidinones with very high yields (> 85 %) and enantioselectivities (> 98 %).

12.5.4 SYNTHESIS OF (R)-N-PROPIONYL-4,5,5-TRIMETHYL-1,3-OXAZOLIDIN-2-ONE

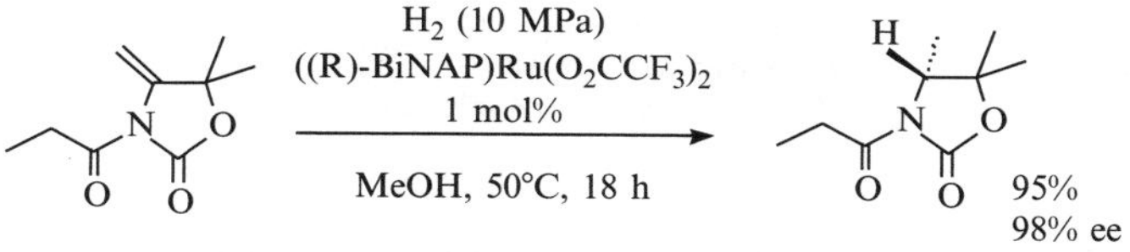

Materials and equipment

- N-Propionyl-5,5-dimethyl-4-methylene-1,3-oxazolidin-2-one [13,14], 0.2 g, 1.2 mmol
- ((R)-BiNAP)Ru(O$_2$CCF$_3$)$_2$[10]. 11 mg, 0.012 mmol
- Dry and degassed methanol, 15 mL

- 125 mL Stainless steel autoclave with mechanical stirrer, thermostatically controlled oven and thermocouple
- 50 mL Round bottomed flask
- Rotary evaporator
- Kugelrohr apparatus

Procedure

1. The 125 mL stainless steel autoclave was flushed with nitrogen.
2. The N-propionyl-5,5-dimethyl-4-methylene-1,3-oxazolidin-2-one, the ruthenium catalyst and methanol (10 mL) were placed in the autoclave under nitrogen atmosphere.
3. The autoclave was sealed, flushed with hydrogen and pressurized with 10 MPa of hydrogen. The mixture was stirred for 18 hours at 50 °C under 10 MPa of hydrogen.
4. Once the autoclave had cooled to room temperature, the autoclave was carefully depressurized, the solution was poured into a 50 mL round bottomed flask and the autoclave was rinsed with methanol (5 mL). The solvent was removed by using a rotary evaporator.
5. The hydrogenated carbamate can be recovered free of ruthenium catalyst by sublimation under reduced pressure using a Kugelrohr apparatus (bp 80 °C, 1.5 mmHg).

    This procedure has been scaled up to provide 1.5 g of 4,4,5-trimethyl-1,3-dioxolane-2-one.

    The optical purity can be determined by using HPLC equipped with a chiral (S,S)-WHELK 0–1 column (250 × 4.6 mm) eluted with a hexane-2-propanol (95/5) mixture.

12.6 ENANTIOSELECTIVE RUTHENIUM-CATALYZED HYDROGENATION OF VINYLPHOSPHONIC ACIDS

VIRGINIE RATOVELOMANANA-VIDAL, JEAN-PIERRE GENÊT

Laboratoire de Synthèse Organique Sélective & Produits Naturels, Ecole Nationale Supérieure de Chimie de Paris, 11, rue Pierre & Marie Curie 75231 Paris cedex 05 France, Tel: 01 44 27 67 43 fax: 01 44 07 10 62, e-mail:genet@ext.jussieu.fr

12.6.1 SYNTHESIS OF CHIRAL RU(II) CATALYSTS

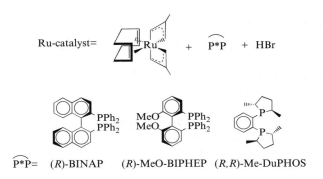

$\widehat{P^*P}=$ (*R*)-BINAP   (*R*)-MeO-BIPHEP   (*R,R*)-Me-DuPHOS

Materials and equipment

- [(COD)Ru(2-(methylallyl)$_2$], 22 mg
- (*R*)-MeO-BIPHEP, 48 mg
- Distilled acetone, 2 mL
- Methanolic hydrobromic acid (0.2 N), 0.72 mL

- Glass tube (10 mL) with a magnetic stirrer bar
- Magnetic stirrer

Procedure

All reactions were carried out under argon in solution in dry solvents.

1. The ruthenium catalyst was prepared at room temperature by reaction of [(COD)Ru(2-(methylallyl)$_2$] with ligand P*P in this case (*R*)-MeO-BIPHEP (1.2 eq) in acetone (2 mL).
2. Methanolic HBr (2.2 eq) was added dropwise to the solution which was subsequently stirred for 30 minutes at room temperature. A yellow precipitate was formed and the solvent was then evaporated *in vacuo*.

12.6.2   ASYMMETRIC HYDROGENATION OF VINYLPHOSPHONIC ACIDS CARRYING A PHENYL SUBSTITUENT AT C_2

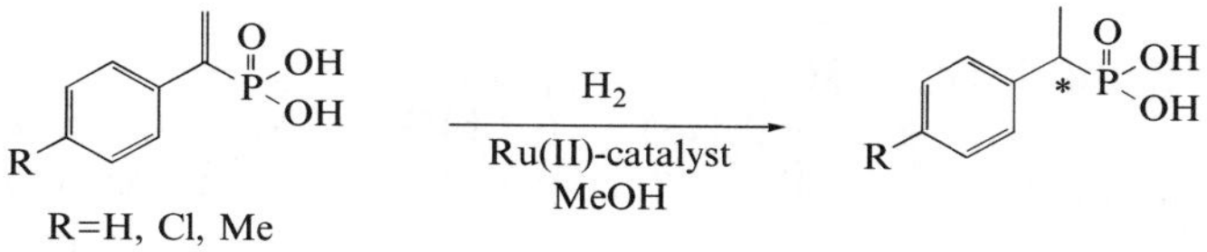

R=H, Cl, Me

Materials and equipment

- Ru(II)-catalyst 1 mol%
- Vinyl phosphonic acid, 1 mmol to 6 mmol
- Methanol, 2 mL to 5 mL

- Autoclave (500 mL)
- Magnetic stirrer

Procedure

All reactions were carried out in solution under argon.

1. A solution of the appropriate substrate (1 mmol or 6 mmol) in degassed methanol (2 mL to 5 mL) was added to the Ru(II) catalyst.
2. The glass vessel was placed under argon in a stainless steel autoclave, which was then pressurized with hydrogen.
3. The reaction proceeded at 10 bar and 80 °C.

    The enantiomeric excesses of the phosphonic acids were measured using [31] P NMR after treatment with (1S, 2S)-(−)-N,N'-dimethyl(diphenylethylene)-diamine in CDCl$_3$ and a catalytic amount of CD$_3$OD.

1-Phenylethenylphosphonic acid (R = H):

^{1}H NMR (200 MHz, CDCl$_3$): δ 9.2 (sl, 2H); 7.3 (s, 5H); 3.0 (qd, J 8.5 Hz, J 24.5 Hz, 1H); 1.5 (dd, J 6.5 Hz, J 18.4 Hz, 3H).
    ^{31}P NMR (101 MHz, CDCl$_3$): δ 35.3 ppm.
    The ee was measured by ^{31}PNMR (101 MHz, CDCl$_3$) in the presence of 1 equivalent of (1S, 2S)-(−)-N, N$'$-diphenyl-ethylenediamine and 4% (vol.) of CD$_3$OD.
    δ (ppm) 26.8 (R,S,S) and 26.4 ppm (S,S,S). Methyl ester [α]$_D$ = +4.5 (c 1.3., CHCl$_3$) for ee 71 % (R).

1-para-Chlorophenylethenylphosphonic acid (R = Cl)

^{1}H NMR (200 MHz, CDCl$_3$): δ8.9 (sl, 2H); 7.3–7.2 (d, *J* 7.6 Hz, 2H); 7.2–7.1 (d, *J* 7.6 Hz, 2H); 3.0 (qd, *J* 7.3 Hz, *J* 23.2 Hz, 1H); 1.4 (dd, *J* 7.2 Hz, *J* 18.4 Hz, 3H).

^{31}P NMR (101 MHz, CDCl$_3$): δ 32.9 ppm.

The ee was measured by ^{31}P NMR (101 MHz, CDCl$_3$) in the presence of 1 equivalent of (1S, 2S)-(−)-N, N'-dimethyl(diphenylethylene)diamine and 1 % (vol.) of CD$_3$OD.

δ (ppm) 26.4 (R,S,S) and 25.9 ppm (S,S,S). Methyl ester [α]$_D$ = 7.0 (*c* 1.3, CHCl$_3$) for ee 80 % (S).

1-para-Methylphenylethenylphosphonic acid (R = Me)

^{1}H NMR (200 MHz, CDCl$_3$): δ 8.5 (sl, 2H); 7.2 (d, *J* 7.8 Hz, 2H); 7.1 (d, *J* 7.8 Hz, 2H); 2.9 (qd, *J* 8.0 Hz, *J* 19.0, 1H); 2.3 (s, 3H); 1.4 (dd, *J* 6.0 Hz, *J* 18.1 Hz, 3H).

^{31}P NMR (101 MHz, CDCl$_3$): δ 35.3 ppm.

The ee was measured by [31] P NMR (101 MHz, CDCl$_3$) in the presence of 1 equivalent of (1S, 2S)-(−)-N,N'-dimethyl-diphenylethylene)diamine and 4 % (vol.) of CD$_3$OD.

δ (ppm) 27.6 (R,S,S) and 27.3 ppm (S,S,S). Methyl ester [α]$_D$ = −7.5 (*c* 1.0, CHCl$_3$) for ee 78 % (S).

12.6.3   ASYMMETRIC REDUCTION OF A VINYLPHOSPHORIC ACID CARRYING A NAPHTHYL SUBSTITUENT AT C$_2$

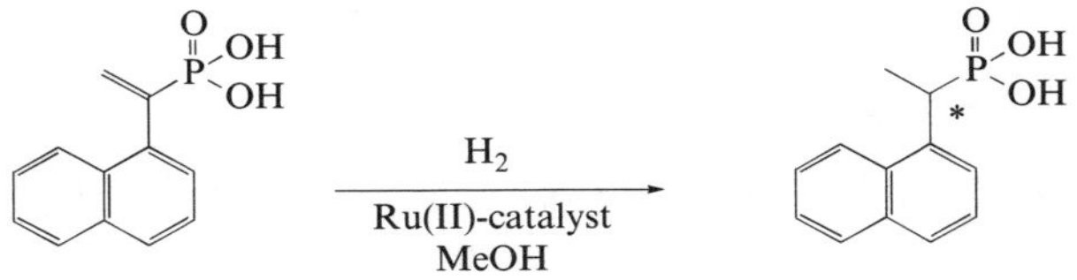

Materials and equipment

- Ru(II)-catalyst, 1 mol %
- Vinyl phosphonic acid, 1 mmol to 6 mmol
- Methanol, 2 mL to 5 mL

- Autoclave (500 mL)
- Magnetic stirrer

Procedure

All reactions were carried out under argon in solution as above. Thus:

1. A solution of the appropriate substrate (1 mmol or 6 mmol) in degassed methanol (2 mL to 5 mL) was added to the Ru(II) catalyst.
2. The glass vessel was placed under argon in a stainless steel autoclave, which was then pressurized with hydrogen.
3. The reaction proceeded at 10 bar and 80 °C.

The enantiomeric excesses of the phosphonic acid was measured using ^{31}P NMR after treatment with (1S, 2S)-(−)-N,N'-dimethyl(diphenyl-ethylene)-diamine in CDCl$_3$ and a catalytic amount of CD$_3$OD.

1-Naphthylethenylphosphonic acid (R = naphthyl)

^{1}H NMR (200 MHz, CDCl$_3$): δ 9.0 (sl, 2H); 7.9–7.8 (m, 2H); 7.74–7.71 (d, J 7.7 Hz, 1H); 7.748–7.34 (m, 4H) 3.7 (qd, J 7.2 Hz, J 25.1, 1H); 1.3 (dd, J 7.1 Hz, J 18.7 Hz, 3H).

^{31}P NMR (101 MHz, CDCl$_3$): δ 32.2 ppm

The ee was measured by ^{31}P NMR (101 MHz, CDCl$_3$) in the presence of 1 equivalent of (*1S, 2S*)-(−)-*N,N'*-dimethyl(diphenylethylene)-diamine and 4% (vol.) of CD$_3$OD.

δ (ppm) 26.7 (*R,S,S*) and 26.3 ppm (*S,S,S*). Methyl ester [α]$_D$ = +94.2 (c 1.0, CHCl$_3$) for ee 86% (*S*).

12.6.4 SCOPE OF THE HYDROGENATION REACTION

The enantioselective ruthenium-catalysed hydrogenation reaction, which is applied above to vinylphosphonic acid derivatives[15], has a much larger scope. It has been shown that a number of olefins and functionalized carbonyl compounds can be hydrogenated with very high selectivity by using the '*in situ*' generated ruthenium catalyst[16]. For instance, β-ketoesters[16], phosphonates[17], sulfides[18], sulfones[19], sulfoxides and β-diketones[20] have been reduced to the corresponding alcohols in enantiomeric excesses approaching 100%. Atropoisomeric ligands (BINAP, BIPHEMP, MeO-BIPHEP) but also DuPHOS, DIOP, SKEWPHOS[21], CnrPHOS[22] etc... can be used as chiral auxiliaries. Selected results are given in the following table. Dynamic kinetic resolution of α-chloro and α-acetamido-β-ketoesters have also been performed by this method, leading to *anti*-α-chloro[23] and *syn*-α-acetamido-β-hydroxyesters[24] in 99% enantiomeric excess.

All these hydrogenation reactions are quantitative, easy to perform on a large scale, and thus represent an highly convenient approach to a number of optically pure compounds. In most cases, it compares favourably with enzyme-promoted reductions which have more limited scope with respect to substrates. The enantioselective hydrogenations have been applied to the synthesis of natural products of biological interest[25].

Table 12.3 Asymmetric ruthenium-catalysed hydrogenations.

Substrate	[Ru]/(P*P)	e.e.(%)	Ref.
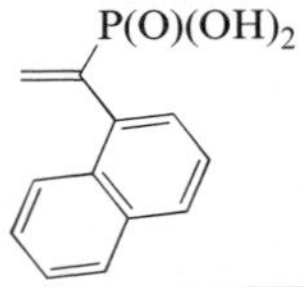 P(O)(OH)$_2$ naphthalene derivative	(R)-MeO-BIPHEP	86(S)	15
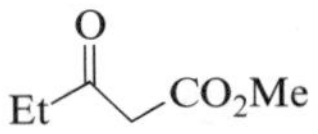 Et—C(O)—CO$_2$Me	(R)-MeO-BIPHEP	99(R)	16
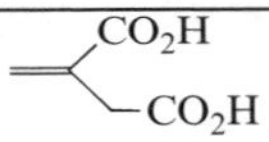 CO$_2$H / CO$_2$H	(R)-BINAP	98(R)	16
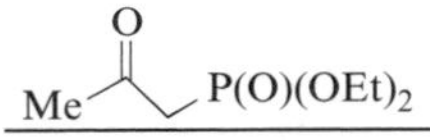 Me—C(O)—P(O)(OEt)$_2$	(R)-BINAP	99(R)	17
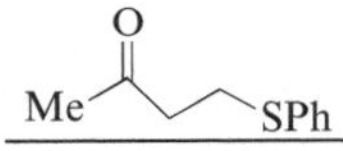 Me—C(O)—SPh	(S)-MeO-BIPHEP	98(S)	18
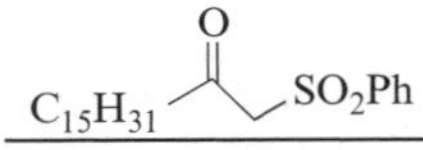 C$_{15}$H$_{31}$—C(O)—SO$_2$Ph	(S)-MeO-BIPHEP	>95(S)	19
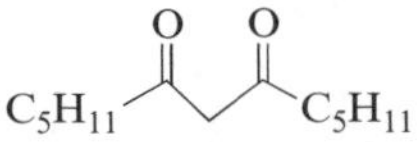 C$_5$H$_{11}$—C(O)—C(O)—C$_5$H$_{11}$	(R)-MeO-BIPHEP	99(R,R) (anti)	20

12.7 SYNTHESIS OF A CYLINDRICALLY CHIRAL DIPHOSPHINE AND ASYMMETRIC HYDROGENATION OF DEHYDROAMINO ACIDS

JAHYO KANG and JUN HEE LEE

Department of Chemistry, Sogang University, Seoul 121–742, Korea

12.7.1 PREPARATION OF (R,R)-1,1′-BIS (α-HYDROXYPROPYL) FERROCENE

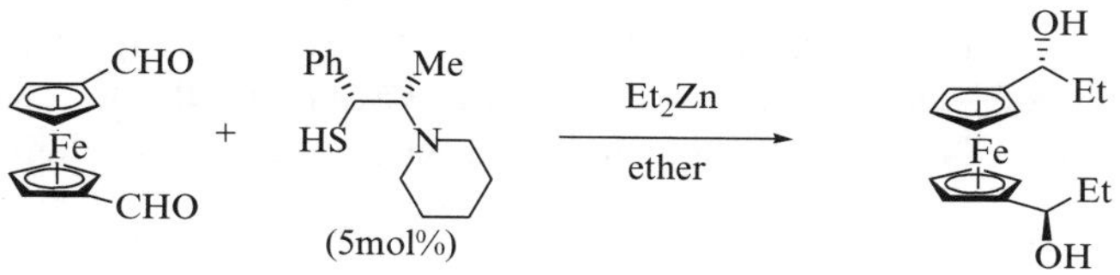

Materials and equipment

- 1,1′-Ferrocenedicarboxaldehyde, 8.30 g[26,27]
- Diethylzinc (1.1 M in toluene), 28.7 mL
- (1R,2S)-1-Phenyl-2-(1-piperidinyl)propane-1-thiol, 400 mg
- Diethyl ether, 115 mL
- 1 M Hydrochloric acid
- Diethyl ether
- Brine
- Magnesium sulfate
- Silica gel (230–400 mesh)

- 250 mL Round-bottomed flask with a magnetic stirring bar
- Magnetic stirrer
- Temperature controller
- Separatory funnel, 250 mL
- Glass filter (3G3)
- Rotatory evaporator
- Glass column

Procedure

1. In a degassed 250 mL round-bottomed flask equipped with a magnetic stirring bar were placed 1,1′-ferrocenedicarboxaldehyde (8.30 g), (1R,2S)-1-phenyl-2-(1-piperidinyl)propane-1-thiol (400 mg) and dry diethyl ether (115 mL). Diethylzinc in toluene (1.1 M, 28.7 mL) was added dropwise to the mixture at 0 °C. The reaction mixture was stirred at 0 °C for 10 hours.
2. After the period, the reaction was quenched by adding a solution of 1 M HCl with vigorous stirring at 0 °C until no more ethane gas was generated. The white inorganic material was removed by filtration over a glass filter.
3. The organic layer was separated and the aqueous layer was extracted with ether (3 × 30 mL). The combined organic layer was washed with brine, dried over magnesium sulfate, filtered and concentrated using a rotatory evaporator to give a crude black residue.

4. The residue was chromatographed on silica gel (eluent: *n*-hexane–ethyl acetate, 3:1) to afford the (*R*,*R*)-1,1'-bis(α-hydroxypropyl)ferrocene as an orange solid (9.83 g, 95%).

   The ee (99.9%) was determined by HPLC (Daicel Chiralcel OJ column, eluent 2-propanol-*n*-hexane 2:98, flow 0.5 mL/min); (*S*,*S*)-enantiomer: R$_t$ 20.0 min, (*R*,*S*)-*meso* isomer: R$_t$ 24.8 min, (*R*,*R*)-enantiomer: R$_t$ 35.6 min; the (*S*,*S*)-isomer was not detected); (*R*,*R*)-isomer: (*R*,*S*)-*meso* isomer = 98.3:1.7.

12.7.2   PREPARATION OF (R,R)-1,1'-BIS[α-(DIMETHYLAMINO) PROPYL]FERROCENE[28]

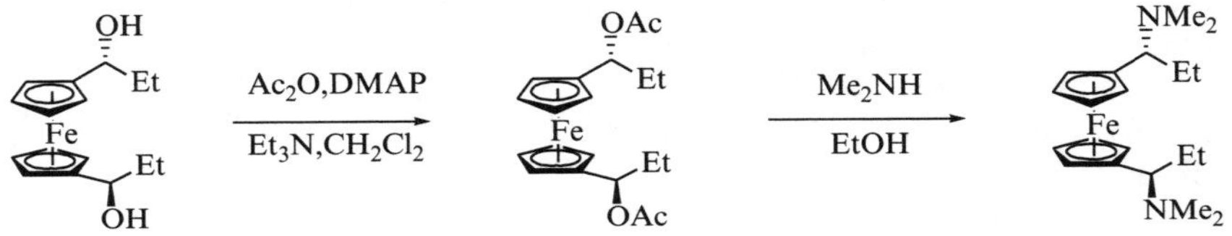

Materials and equipment

- Dichloromethane, 54 mL
- 4-Dimethylaminopyridine
- Triethylamine, 36.4 mL
- Acetic anhydride, 12.2 mL
- 50% Aqueous dimethylamine, 22.2 mL
- Absolute ethyl alcohol, 90 mL
- 10% Phosphoric acid solution
- 10% Sodium hydroxide solution
- Diethyl ether
- Magnesium sulfate, potassium carbonate

- 250 mL Round-bottomed flask with a magnetic stirring bar
- Magnetic stirrer
- Separatory funnel, 250 mL, 1 L
- Rotatory evaporator

Procedure

1. In a 250 mL round-bottomed flask equipped with a magnetic stirring bar were placed dichloromethane (54 mL), (*R*,*R*)-1,1'-bis(α-hydroxypropyl)ferrocene (9.84 g) and a catalytic amount of 4-(dimethylamino)pyridine (39.8 mg) under nitrogen. Triethylamine (36.4 mL) and acetic anhydride (12.2 mL) were

added to the mixture successively at 0 °C, and the resulting mixture was stirred at room temperature for 8 hours.

2. Cold water (ice–water, 50 mL) was added, and the mixture was extracted with dichloromethane (3 × 30 mL). The combined extracts were dried over magnesium sulfate, filtered and concentrated using a rotatory evaporator to afford the diacetate as a dark-brown residue.

3. To the residue in a 250 mL round-bottomed flask equipped with a magnetic stirring bar were added 50 % aqueous dimethylamine (22.2 mL) and absolute ethyl alcohol (90 mL), the mixture was stirred at room temperature for 48 hours. During this time, an orange solid was formed.

4. The solvent was removed using a rotatory evaporator, and the resulting residue was diluted with ether (50 mL). The diamine was extracted with 10 % phosphoric acid (3 × 15 mL), after which the aqueous layer was made alkaline (pH 9) with 10 % sodium hydroxide solution (100 mL). The resulting mixture was extracted with ether (5 × 50 mL). The combined ethereal extracts were dried over anhydrous potassium carbonate, filtered and concentrated *in vacuo* to afford the pure (*R,R*)-1,1′-bis [α-(dimethylamino) propyl]ferrocene as an orange solid (11.4 g, 98.0 %).

12.7.3   PREPARATION OF (R,R,$_p$S,$_p$S)-1,1′-BIS [α-(DIMETHYLAMINO) PROPYL]-2,2′-BIS(DIPHENYL-PHOSPHINO)FERROCENE

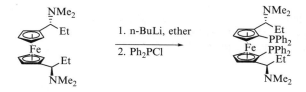

Materials and equipment

- Diethyl ether, 22 mL
- *n*-BuLi (1.68 M in hexanes), 21.3 mL
- Chlorodiphenylphosphine, 10.3 mL
- Saturated sodium bicarbonate solution
- Diethyl ether
- Brine
- Magnesium sulfate
- Silica gel (230–400 mesh)

- 250 mL Round-bottomed flask with a magnetic stirring bar
- Magnetic stirrer
- Syringe pump

- Separatory funnel, 250 mL
- Rotatory evaporator

Procedure

1. (R,R)-1,1′ -Bis[α-(dimethylamino)propyl]ferrocene (5.10 g) was placed in a 250 mL round-bottomed flask equipped with a magnetic stirring bar under nitrogen; dry diethyl ether (22 mL) was then added. To the mixture was added dropwise n-BuLi in hexanes (1.68 M, 34.0 mL) within 10 minutes at room temperature. After 30 minutes the colour of the mixture changed from yellow to red.
2. After 6 hours, chlorodiphenylphosphine (18 mL) was added over 2 hours with the help of a syringe pump. After the addition was complete, the resulting suspension was stirred at room temperature for 3 hours and aqueous sodium bicarbonate was slowly added to hydrolyse the excess chlorodiphenylphosphine with cooling in an ice bath.
3. The reaction mixture was extracted with diethyl ether (3 × 30 mL). The combined organic layer was washed with brine, dried over magnesium sulfate, filtered and concentrated using a rotatory evaporator. The resulting residue was chromatographed (eluent: n-hexane–ethyl acetate, 97:3) on silica gel pre-deactivated with triethylamine: n-hexane (2:98) to afford the $(R, R, _pS, _pS)$-1,1′-bis[α-(dimethylamino)propyl]-2,2′-bis(diphenylphosphino)ferrocene as an orange solid (8.07 g, 78.0 %).

12.7.4   PREPARATION OF $(R, R, _pS, _pS)$-1,1′-BIS (α-ACETOXYPROPYL)-2,2′-BIS (DIPHENYL-PHOSPHINO)FERROCENE[29]

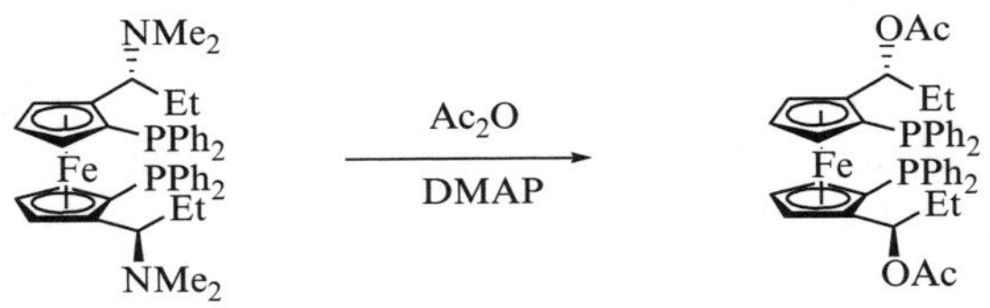

Materials and equipment

- Acetic anhydride, 4.00 mL
- 4-Dimethylaminopyridine
- Silica gel (230–400 mesh)

- 25 mL Schlenk-type flask with a magnetic stirring bar
- Magnetic stirrer
- Rotatory evaporator

Procedure

1. $(R, R, {}_pS, {}_pS)$-1,1'-bis [α-(Dimethylamino)propyl]-2,2'-bis(diphenylphosphino)ferrocene (1.55 g) and 4-dimethylaminopyridine were placed in a 25 mL degassed Schlenk flask equipped with a magnetic stirring bar; acetic anhydride (4.00 mL) was then added.
2. The reaction mixture was heated at 100 °C for 20 hours, after which time the excess acetic anhydride was removed under high vacuum at 50 °C.
3. The resulting residue was chromatographed (eluent: n-hexane–ethyl acetate, 95:5) on silica gel pre-deactivated with triethylamine: n-hexane (2:98) to afford the $(R, R, {}_pS, {}_pS)$-1,1'-bis(α-acetoxypropyl)-2,2'-bis(diphenylphosphino)ferrocene as an orange solid (1.26 g, 78.0%).

12.7.5   PREPARATION OF $({}_pS, {}_pS)$-1,1'-BIS(DIPHENYLPHOSPHINO)-2,2'-BIS(1-ETHYLPROPYL)FERROCENE [(S,S)-3-PT-FERROPHOS]

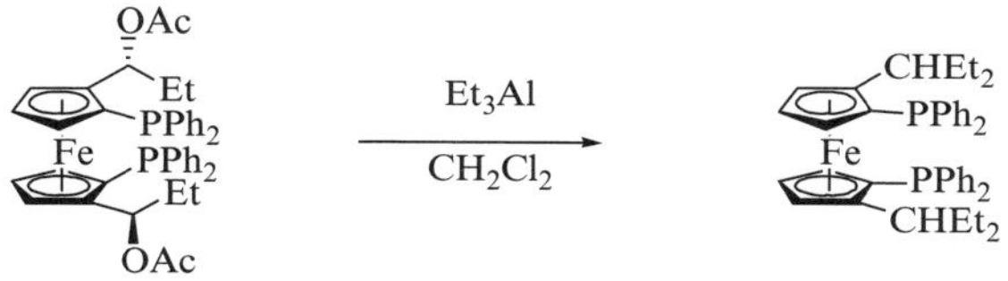

Materials and equipment

- Dichloromethane, 16.6 mL
- Triethylaluminium (1.35 M in toluene), 6.14 mL
- Saturated sodium bicarbonate solution
- Saturated sodium potassium tartrate solution
- Diethyl ether, 30 mL
- 1 M Hydrochloric acid
- Brine
- Magnesium sulfate

- 100 mL Round-bottomed flask with a magnetic stirring bar
- Magnetic stirrer
- Separatory funnel, 125 mL
- Rotatory evaporator

Procedure

1. In a 100 mL round-bottomed flask equipped with a magnetic stirring bar were placed dichloromethane (16.6 mL) and $(R, R, {}_pS, {}_pS)$-1,1'-bis (α-acetoxypropyl)-2,2'-bis(diphenylphosphino)ferrocene (1.25 g) under nitrogen. Triethylaluminium in toluene (1.35 M, 6.14 mL) was added to the mixture

at $-20\,°C$. The cold bath was removed immediately and the reaction mixture was warmed to room temperature.

2. After stirring at room temperature for 20 minutes, the reaction mixture was cooled to $0\,°C$ and transferred via a cannula into saturated aqueous $NaHCO_3$ (15 mL). Saturated aqueous sodium potassium tartrate was added (15 mL).

3. The dichloromethane was removed by rotatory evaporation and dry ether (30 mL) was added. The mixture was stirred vigorously for 15 minutes and then acidified with 1 N HCl (15 mL).

4. The organic layer was separated, and the aqueous layer was extracted with ether (3 × 20 mL). The combined organic layer was washed with brine, dried over magnesium sulfate, filtered and concentrated using a rotatory evaporator. The crude orange solid was recrystallized from hot ethanol to give the ($_pS$, $_pS$)-1,1′-bis(diphenylphosphino)-2,2′-di-γ-pentylferrocene   [(S,S)-3-Pt-FerroPHOS] as yellow crystals (519 mg, 45.0 %).

12.7.6   PREPARATION OF [(COD)RH(($_pS$, $_pS$)-1,1′-BIS (DIPHENYLPHOSPHINO)-2,2′-BIS(1-ETHYLPROPYL) FERROCENE)]$^+ BF_4^-$

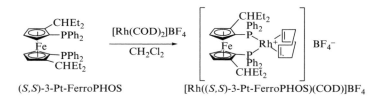

(S,S)-3-Pt-FerroPHOS   [Rh((S,S)-3-Pt-FerroPHOS)(COD)]BF₄

Materials and equipment

- Dichloromethane
- Bis(1,5-cyclooctadiene)rhodium(I) tetrafluoroborate, 111 mg

- 25 mL Schlenk-type flask with a magnetic stirring bar
- 10 mL pressure-equalized dropping funnel
- Magnetic stirrer
- Glass filter (3G4)

Procedure

1. In a 25 mL Schlenk-type flask, equipped with a 10 mL pressure equalized dropping funnel and a magnetic stirring bar, were placed bis(1,5-cyclooctadiene)rhodium(I) tetrafluoroborate (111 mg) and dichloromethane (5 mL) under nitrogen.

2. A solution of (S,S)-3-Pt-FerroPHOS (200 mg) in dichloromethane (5 mL) was added dropwise over 20 minutes to the above mixture from the dropping funnel at 23 °C. The resulting deep red homogeneous solution was stirred for 3 hours at that temperature.

3. During this time, a red orange coloured solution was formed. The solvent was carefully removed by a stream of nitrogen gas to afford a gold creamy solid. The creamy solid was recrystallized from $CHCl_3/Et_2O$, filtered, washed with ether and dried *in vacuo* to afford a Rh complex as orange crystals (230 mg, 84.6%).

12.7.7 ASYMMETRIC HYDROGENATION OF α-ACETAMIDO CINNAMIC ACID

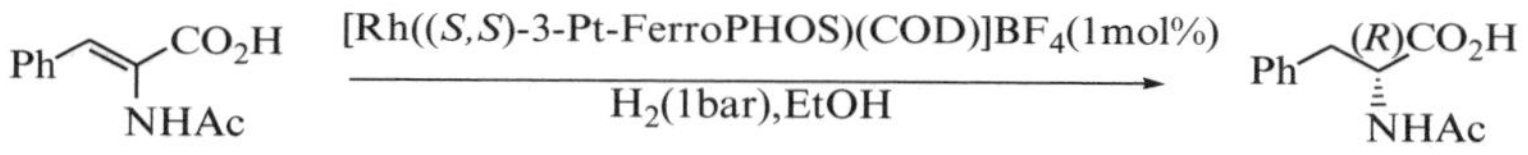

Materials and equipment

- {Rh(COD)[(S,S)-3-Pt-FerroPHOS]}BF_4, 9.9 mg
- Absolute ethyl alcohol, 3.3 mL
- α-Acetamido cinnamic acid, 205 mg
- Degassed ethanol (3.3 mL)

- Dry box
- 25 mL Schlenk-type flask with a magnetic stirring bar
- Magnetic stirrer

Procedure

1. In a dry box, a 25 mL Schlenk-type flask with a magnetic stirring bar was charged with {Rh(COD)[(S,S)-3-Pt-FerroPHOS]}BF_4 (9.9 mg), α-acetamido cinnamic acid (205 mg) and degassed ethanol (3.3 mL). After sealing, the flask was removed from the dry box.

2. With stirring of the mixture at 20–23 °C, the flask was then freeze–pump–thaw–degassed (3 cycles) and stirred under hydrogen gas (*ca.* 1 bar) for 2 hours.

The enantiomeric excess (98.9%) was determined by GC (Chrompack Chiralsil-L-Val column)

The reaction depends on various factors including solvent, initial pressure, catalyst precursor and the N-protecting group. Due to the high stability of (S,S)-3-Pt-FerroPHOS towards air, it may be used in an industrial process.

Conclusion

The cylindrically chiral diphosphine neither changed, nor lost its reactivity and selectivity in hydrogenation reactions even after long exposure to air. In a ^{31}P-NMR study, no detectable air-oxidation was observed even after a long exposure (3 weeks) to the atmosphere. The procedures for the synthesis of the chiral ligand and the asymmetric reaction described above are very simple, giving high enantioselectivity with many dehydroamino acids (Table 12.4).

Table 12.4 Rh-Catalyzed asymmetric hydrogenation of α-(acylamino)acrylic acids and esters.

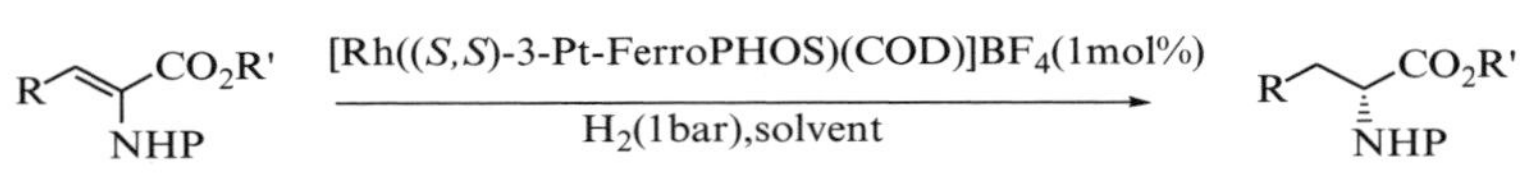

R	R'	P	Solvent	% ee	Config
Ph	H	Ac	EtOH	98.7	**R**
Ph	H	Ac	EtOH	98.9	R
Ph	Me	Ac	EtOH	97.6	R
Ph	Me	Cbz	MeOH	85.3	R
H	H	Ac	MeOH	98.2	R
H	Me	Ac	MeOH	97.5	R
2-Np	H	Ac	MeOH	95.7	R

12.8 SYNTHESIS AND APPLICATION OF DIAMINO FERRIPHOS AS A LIGAND FOR ENANTIOSELECTIVE Rh-CATALYSED PREPARATION OF CHIRAL α-AMINO ACIDS

MATTHIAS LOTZ[a], JUAN J. ALMENA PEREA[b] and PAUL KNOCHEL[a]

[a]Institut für Organische Chemie, Ludwig-Maximilians-Universität München Butenandtstr. 5-13, D-81377 München, Germany
[b]Degussa-Hüls AG, Fine Chemicals Division, Rodenbacher Chaussee 4, D-63403 Hanau, (Wolfgang), Germany

12.8.1 SYNTHESIS OF 1,1'-DI(BENZOYL)FERROCENE

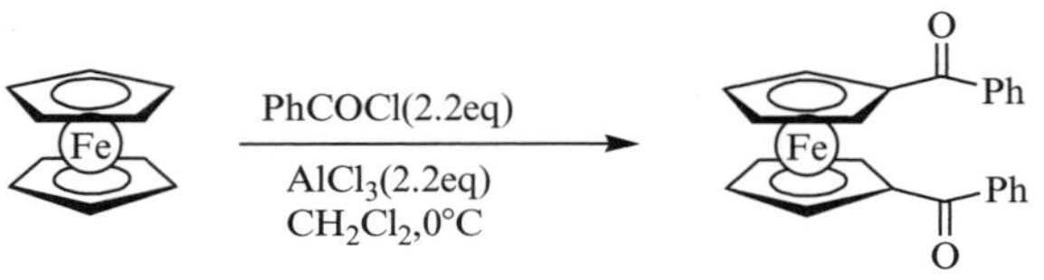

Materials and equipment

- Ferrocene, 9.30 g, 50.0 mmol
- Benzoyl chloride, 12.77 mL, 110.0 mmol
- Aluminum (III) chloride, 14.67 g, 110.0 mmol
- Dry dichloromethane, 210 mL
- Water
- Brine
- Saturated potassium carbonate solution
- *n*-Pentane, *t*-butyl methyl ether
- Silica gel (60, 0.063–0.0200 mm, 70–200 mesh ASTM, Merck)

- 250 mL Two neck round-bottomed flask with an argon inlet
- 250 mL Dropping funnel
- Magnetic stirring bar
- Magnetic stirrer
- Syringe (20 mL)
- Separatory funnel, 500 mL
- Chromatography column

Procedure

1. In a 250 mL two-necked round-bottomed flask with an argon inlet, equipped with a dropping funnel and a magnetic stirring bar, aluminium (III) chloride (14.67 g) was suspended under argon in dichloromethane (60 mL) and cooled to 0 °C in an ice bath. Via a syringe, benzoyl chloride (12.77 mL) was added. Ferrocene (9.30 g), dissolved in dichloromethane (50 mL), was added dropwise within 30 minutes. The reaction was allowed to warm to room temperature and stirred overnight.
2. The work-up was done by dropwise addition of ice-cold water (50 mL; **caution: gas evolution!**). The reaction mixture was diluted with dichloromethane (100 mL) and washed with saturated potassium carbonate solution (2 × 50 mL) and brine (2 × 50 mL). The organic layer was dried over magnesium sulfate, filtrated and concentrated using a rotatory evaporator.
3. The crude product was purified by column chromatography (*n*-pentane: *t*-butyl methyl ether: dichloromethane 5:4:1) yielding 1,1′-di(benzoyl)ferrocene (16.16 g, 41.0 mmol, 82%) as a dark red solid (mp 97–100 °C).

   ^{1}H-NMR (200 MHz, CDCl$_3$): δ 7.77–7.72 (m, 4 H), 7.50–7.38 (m, 6 H), 4.88 (t, *J* 1.8 Hz, 4 H), 4.53 (t, *J* 1.8 Hz, 4 H).

   ^{13}C-NMR (50 MHz, CDCl$_3$): δ 197.71, 138.94, 131.76, 128.18, 127.95, 79.36, 74.46, 72.95.

12.8.2   SYNTHESIS OF (S,S)-1,1′-BIS (α-HYDROXYPHENYLMETHYL)FERROCENE

Materials and equipment

- 1,1′Di(benzoyl)ferrocene, 3.00 g, 11.1 mmol
- CBS-catalyst[30]: B-methyl oxazaborolidine (prepared from (R)-2-(diphenyl-hydroxy-methyl)pyrrolidine and methyl boronic acid)[31], 1.85 g, 6.7 mmol
- Borane dimethyl sulfide-complex, 2.00 mL, 21.1 mmol
- Dry tetrahydrofuran, 50 mL
- Methanol, 3 mL
- Saturated ammonium chloride solution, 150 mL
- t-Butyl methyl ether
- Diethyl ether
- Brine
- Magnesium stirrer
- Silica gel (60, 0.063–0.0200 mm, 70–200 mesh ASTM, Merck)

- 250 mL Round-bottomed flask with an argon inlet
- Magnetic stirring bar
- Magnetic stirrer
- Ice bath
- Two syringes (20 mL)
- Separatory funnel, 500 mL
- Rotatory evaporator
- Chromatography column

Procedure

1. In a 250 mL round-bottomed flask with an argon inlet equipped with a magnetic stirring bar the CBS-catalyst (1.85 g) was dissolved in tetrahydrofuran (10 mL) and cooled to 0 °C in an ice bath. From a syringe filled with borane dimethyl sulfide-complex (2.00 mL dissolved in 10 mL THF) 20 % of the volume (2.40 mL) were added and the solution was stirred for 5 minutes. A solution of the diketone (3.00 g dissolved in 30 mL THF) was added from a second syringe simultaneously with the rest of the borane dimethyl sulfide-complex over 2 hours. The resulting yellow solution was stirred for another

10 minutes at 0 °C and the excess borane dimethyl sulfide-complex was destroyed by dropwise addition of methanol (**caution: gas evolution!**).
2. After no further gas evolution could be detected, the reaction mixture was poured into saturated ammonium chloride solution (150 mL) and transferred into a separatory funnel. The aqueous layer was extracted with *t*-butyl methyl ether (3 × 70 mL), the combined organic layers were washed with water (2 × 100 mL) and brine (100 mL) and dried over magnesium sulfate. After filtration the solvent was removed using a rotatory evaporator (bath temperature <30 °C) to give a yellow oil.
3. The crude product was purified by column chromatography (*n*-pentane: diethyl ether 1:1) and dried under vacuum yielding (*S*,*S*)-1,1′-bis(α-hydroxyphenyl-methyl)ferrocene (2.93 g, 10.7 mmol, 96%) as a yellow solid (mp 128–130 °C).

The ee (> 99%) was determined by HPLC (Daicel® OD column, flow 0.6 mL/min, 215 nm, eluent 2-propanol/*n*-heptane 5/95); (*SS* and *RS*): R_t 26.53 min, (*RR*): R_t 30.70 min.

^{1}H-NMR (300 MHz, CDCl$_3$): δ 7.27–7.17 (m, 10 H), 5.45 (s, br, 4 H), 4.42 (s, br, 2 H), 4.22 (s, br, 2 H), 4.16 (s, br, 2 H), 4.11 (s, br, 2 H).

^{13}C-NMR (75 MHz, CDCl$_3$): δ 144.08, 128.17, 127.40, 126.19, 93.45, 72.59, 68.10, 67.89, 66.70, 66.66.

12.8.3   SYNTHESIS OF (*S*,*S*)-1,1′-BIS (α-ACETOXYPHENYLMETHYL)FERROCENE

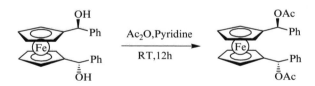

Materials and equipment

- (*S*,*S*)-1,1′-Bis (α-hydroxyphenylmethyl)ferrocene, 2.50 g, 6.3 mmol
- Acetic anhydride, 2 mL
- Pyridine, 4 mL

- 50 mL Round-bottomed flask
- Magnetic stirring bar
- Magnetic stirrer

Procedure

1. In a 50 mL round-bottomed flask equipped with a magnetic stirring bar (*S*,*S*)-1,1′-bis(α-hydroxyphenylmethyl)ferrocene (2.50 g) was dissolved in

pyridine (4 mL) and acetic anhydride (2 mL) was added. The reaction mixture was stirred for 12 hours at room temperature.

2. All volatile material was evaporated under vacuum (1 mmHg, 3 hours) yielding (S,S)-1,1'-bis(α-acetoxyphenylmethyl)ferrocene in quantitative yield as a brown oil.

^{1}H-NMR (300 MHz, CDCl$_3$): δ 7.30–7.20 (m, 10 H), 6.57 (s, 2 H), 4.25–4.23 (m, 2 H), 4.03–4.02 (m, 2 H), 3.98–3.97 (m, 2 H), 3.85–3.84 (m, 2 H), 2.04 (s, 6 H).

^{13}C-NMR (75 MHz, CDCl$_3$): δ 169.91, 139.97, 128.27, 128.03, 127.17, 88.46, 74.06, 69.32, 69.27, 68.58, 68.37, 21.26.

12.8.4   SYNTHESIS OF (S,S)-1,1'-BIS(α-N,N-DIMETHYLAMINOPHENYLMETHYL)FERROCENE

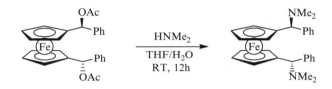

Materials and equipment

- (S,S)-1,1'-Bis(α-acetoxyphenylmethyl)ferrocene, 3.00 g, 6.2 mmol
- Tetrahydrofuran
- Distilled water
- Dimethylamine (40% in water), 10 mL
- t-Butyl methyl ether
- Diethyl ether
- Brine
- Magnesium sulfate
- n-Pentane
- Triethylamine
- Silica gel (60, 0.063–0.0200 mm, 70–200 mesh ASTM, Merck)

- 50 mL Round-bottomed flask
- Magnetic stirring bar
- Magnetic stirrer
- Chromatography column

Procedure

1. In a 50 mL round-bottomed flask equipped with a magnetic stirring bar (S, S)-1,1'-bis(α-acetoxyphenylmethyl)ferrocene (3.00 g) was dissolved in tetra-

hydrofuran and dimethylamine (10 mL, 40% in water) was added. Then water was added dropwise until a yellow solid started to precipitate, whereupon the solid was dissolved again by addition of tetrahydrofuran and the suspension was stirred for 12 hours at room temperature.

2. The tetrahydrofuran was removed under vacuum (1 mmHg), water was added (50 mL) and the solution was transferred into a separatory funnel. After extraction with *t*-butyl methyl ether (3 × 100 mL) the combined organic layers were washed with water (2 × 50 mL) and brine (2 × 50 mL) and dried over magnesium sulfate. After filtration the solvent was removed using a rotatory evaporator to give a yellow oil.

3. The crude product was purified by column chromatography (*n*-pentane: diethyl ether 3:1, 1% triethylamine) and dried under vacuum yielding (*S, S*)-1,1'-bis(α-*N,N*-dimethylaminophenylmethyl)ferrocene (2.45 g, 5.41 mmol, 87%) as a brown solid (48–49 °C).

^{1}H-NMR (300 MHz, CDCl$_3$): δ 7.43–7.28 (m, 10 H), 3.89–3.88 (m, 2 H), 3.61 (s, br, 2 H), 3.56–3.55 (m, 2 H), 3.50–3.49 (m, 4 H), 1.98 (s, 12 H).

^{13}C-NMR (75 MHz, CDCl$_3$): δ 143.34, 128.40, 127.97, 127.00, 90.38, 72.39, 71,39, 70.07, 67.74, 67.67, 44.48.

12.8.5   SYNTHESIS OF (αS, α'S)-1,1'-BIS (α-N, N-DIMETHYLAMINOPHENYLMETHYL)-(R,R)-1,1'-BIS(DIPHENYLPHOSPHINO)FERROCENE

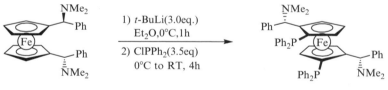

1) *t*-BuLi(3.0eq.)
   Et$_2$O,0°C,1h

2) ClPPh$_2$(3.5eq)
   0°C to RT, 4h

diamino FERRIPHOS

Materials and equipment

- (*S,S*)-1,1'-Bis(α-*N,N*-dimethylaminophenylmethyl)ferrocene, 480 mg, 1.06 mmol
- *t*-BuLi (1.60 M in pentane), 1.99 mL, 3.18 mmol
- Chlorodiphenylphosphine, 0.67 mL, 3.71 mmol
- Dry diethyl ether
- *n*-Pentane
- Saturated sodium hydrogen carbonate solution
- Magnesium sulfate
- Silica gel (60, 0.063–0.0200 mm, 70–200 mesh ASTM, Merck)

- 100 mL Round-bottomed flask with an argon inlet
- Two syringes (5 mL, 1 mL)
- Separatory funnel, 500 mL
- Magnetic stirring bar
- Magnetic stirrer
- Rotatory evaporator
- Chromatography column

Procedure

1. In a 100 mL round-bottomed flask with an argon inlet equipped with a magnetic stirring bar (S,S)-1,1'-bis(α-N,N-dimethylaminophenylmethyl)ferrocene (480 mg) was dissolved in diethyl ether (20 mL) under argon and cooled in an ice bath to 0 °C. t-BuLi (1.60 M in pentane, 1.99 mL) was added within 10 minutes via a syringe (after a few minutes the colour of the solution turned from yellow to dark red). After 1 hour of stirring, chlorodiphenylphosphine (0.67 mL) was added at 0 °C via a syringe and the resulting mixture was stirred for 4 hours at room temperature.
2. After addition of saturated sodium hydrogen carbonate solution (20 mL) the organic layer was separated and the aqueous layer extracted with diethyl ether (3 × 70 mL). The combined organic layers were dried over magnesium sulfate, filtrated and the solvent removed using a rotatory evaporator to give a yellow oil.
3. The crude product was purified by column chromatography (n-pentane: diethyl ether 1:1) immediately after isolation and dried under vacuum yielding (αS, $\alpha' S$)-1,1'-bis(α-N,N-dimethylaminophenyl-methyl)-(R,R)-1,1'-bis(diphenylphosphino)-ferrocene (392 mg, 0.48 mmol, 45%) as a yellow solid (mp 245–246 °C).

   It is important that the crude product is purified as quickly as possible, because in the crude reaction mixture it tends to become oxidized and/or degrade with time.

   ^{1}H-NMR (300 MHz, CDCl$_3$): δ 7.35–7.10 (m, 30 H), 4.52 (s, br, 2 H), 4.39 (s, br, 2 H), 3.29 (s, br, 2 H), 3.15 (s, br, 2 H), 1.51 (s, 12 H).

   ^{13}C-NMR (75 MHz, CDCl$_3$): δ 139.99, 139.68 (d, J 6.8 Hz), 137.84 (d, J 10.1 Hz), 134.77 (d, J 23.0 Hz), 132.38 (d, J 13.4 Hz), 128.55, 128.49, 127.97 (d, J 8.0 Hz), 127.92, 127.44 (d, J 7.0 Hz), 127.30, 126.59, 98.09 (d, J 22.5 Hz), 76.51 (d, J 10.0 Hz), 73.13, 72.88 (d, J 5.2 Hz), 71.57, 68.27 (d, J 10.1 Hz), 42.00.

   ^{31}P-NMR (81 MHz, CDCl$_3$): δ − 23.89.

12.8.6 ASYMMETRIC HYDROGENATION OF METHYL-(Z)-3-PHENYL-2-METHYL-CARBOXAMIDO-2-PROPENOATE USING (S)-(R)-DIAMINO FERRIPHOS AS THE CHIRAL LIGAND

Materials and equipment

- $Rh(COD)_2BF_4$, 4.1 mg, 0.01 mmol
- $(\alpha S, \alpha' S)$-1,1'-Bis(α-N,N-dimethylaminophenyl-methyl)-(R,R)-1,1'-bis(diphenylphosphino)ferrocene, 8.2 mg, 0.01 mmol
- Methyl-(Z)-3-phenyl-2-methylcarboxamido-2-propenoate, 219 mg, 1.0 mmol
- Dry methanol, 9 mL
- Dry toluene, 1 mL
- t-Butyl methyl ether
- Silica gel (60, 0.063–0.0200 mm, 70–200 mesh ASTM, Merck)

- 25 mL Schlenk tube
- Magnetic stirring bar
- Magnetic stirrer
- Balloon, filled with hydrogen
- Chromatography column

Procedure

1. In a 25 mL Schlenk tube equipped with a magnetic stirring bar, under argon, $Rh(COD)_2BF_4$ (4.1 mg) and the ferrocenyl ligand (8.2 mg) were dissolved in methanol/toluene (5:1, 5 mL). After complete solubilization of the rhodium complex, the α-acetamido acrylate (219 mg) was added (dissolved in 5 mL methanol). The Schlenk tube was connected to a hydrogen balloon and the inert gas atmosphere was replaced by hydrogen.
2. The reaction was monitored by ^{1}H-NMR. When complete conversion was obtained, the solvent was removed and the crude reaction was filtered through a short silica gel column using t-butyl methyl ether as eluent. The resulting solution was concentrated using a rotatory evaporator to give N-acetylphenylalanine methyl ester in quantitative yield as a white solid.

    The enantiomeric excess (97.5%) was determined by GC (25 m × 0.2 mm fused silica WCOT Chirasil-L-Val (0.12 μm) using hydrogen (100 kPa) as the mobile phase, 140 °C); (R): R_t 10.13 min, (S): R_t 11.67 min.

^{1}H-NMR (300 MHz, CDCl$_3$): δ 7.25–7.18 (m, 3H), 7.04–7.00 (m, 2H), 5.96 (d, J 7.1 Hz, 1 H), 4.85–4.78 (m, 1 H), 3.65 (s, 3 H), 3.11–2.97 (m, 2H), 1.90 (s, 3 H).

^{13}C-NMR (75 MHz, CDCl$_3$): δ 172.07, 169.52, 135.85, 129.18, 128.51, 127.06, 53.10, 52.21, 37.83, 23.02.

Conclusion

The straightforward synthesis of the diamino FERRIPHOS ligand offers a convenient access to this class of ferrocenyl ligands[32] and makes this ligand well suited for applications in asymmetric hydrogenation.

Table 12.5 shows some examples of α-acetamidoacrylates that were hydrogenated with (S)-(R)-diamino FERRIPHOS as ligand[33].

Table 12.5 Asymmetric hydrogenation of α-acetamidoacrylates using the (S)-(R)-diamino FERRIPHOS ligand.

Substrate	Conversion [%]	ee [%]
R = H	100	97.8 (S)
R = Ph	100	97.5 (S)
R = 2-Naphthyl	100	97.7 (S)
R = p-Cl-Ph	100	98.7 (S)
R = p-F-Ph	100	97.2 (S)

Furthermore FERRIPHOS ligands bearing alkyl groups instead of dimethylamino substituents proved to be excellent ligands in the asymmetric hydrogenation of α-acetamidoacrylic acids[34] and acetoxy acrylic esters[35]. Their air stability and the easy modification of their structure make the FERRIPHOS ligands particularly useful tools for asymmetric catalysis.

REFERENCES

1. Burk, M.J., Fenf, S., Gross, M.F., Tumas, W. *J. Am. Chem. Soc.*, 1995, **117**, 8277.
2. Burk, M.J., Gross, M.F., Martinez, J.P. *J. Am. Chem. Soc.*, 1995, **117**, 9375.
3. Burk, M.J., Feaster, J.E., Nugent, W.A., Harlow, R.L. *J. Am. Chem. Soc.*, 1993, **115**, 10125.
4. Noyori, R. *Acta Chemica Scandinavica* 1996, **50**, 380.
5. Leonard, J., Lygo, B., Procter, G. *Advanced Practical Organic Chemistry*; Second Edition ed., Blackie Academic and Professional;, 1995.

6. Burk, M.J., Feaster, J.E., Harlow, R.L. *Tetrahedron: Asymmetry*, 1991, **2**, 569.
7. Derrien, N., Dousson, C.B, Roberts, S.M., Berens, U., Burk, M.J., Ohff, M., *Tetrahedron: Asymmetry*, 1999, **10**, 3341.
8. Le Gendre, P., Braun, T., Bruneau, C., Dixneuf, P.H. *J. Org. Chem.*, 1996, **61**, 8453.
9. (a) Fournier, J., Bruneau, C., Dixneuf, P.H. *Tetrahedron Lett.*, 1989, **30**, 3981. (b) Joumier, J.M., Fournier, J., Bruneau, C., Dixneuf, P.H., *J. Chem. Soc., Perkin Trans. 1*, 1991, 3271.
10. (a) Doucet, H., Le Gendre, P., Bruneau, C., Dixneuf, P.H., *Tetrahedron: Asymmetry*, 1996, **7**, 525. (b) Genêt, J.P., Pinel, C., Ratovelomanana-Vidal, V., Mallart, S., Pfister, X., Bischoff, L., Cano De Andrade, M.C., Darses, S., Galopin, C., Laffite, J.A. *Tetrahedron: Asymmetry*, 1994, **5**, 675. (c) T. Ohta, H. Takaya, R. Noyori, *Inorg. Chem.*, 1988, **27**, 566. (d) Heiser, B., Broger, E.A., Crameri, Y. *Tetrahedron: Asymmetry*, 1991, **2**, 51.
11. Schurig, V., Betschinger, F. *Bull. Soc. Chim. Fr.*, 1994, **131**, 555.
12. Ager, D.J., Prakash, I., Schaad, D.R. *Chem. Rev.*, 1996, **96**, 835.
13. Le Gendre, P., Thominot, P., Bruneau, C. , Dixneuf, P.H. *J. Org. Chem.*, 1998, **63**, 1806.
14. Bruneau, C., Dixneuf, P.H. *J. Mol. Catal*, 1992, **74**, 97.
15. Henry, J.C., Lavergne, D., Ratovelomanana-Vidal, V., Beletskaya; I.P., Dolgina, T.M. *Tetrahedron Lett.*, 1998, **39**, 3473–6.
16. Genêt, J.-P., Ratovelomanana-Vidal, V., Caño de Andrade, M.C., Pfister, X., Guerreiro, P., Lenoir, J.Y. *Tetrahedron Lett.*, 1995, **36**, 4801–4. Genêt, J.-P., Caño de Andrade, M.C., Ratovelomanana-Vidal, V. *Tetrahedron Lett.*, 1995, **36**, 2063–6. Genêt, J.-P., Pinel, C., Ratovelomanana-Vidal, V., Mallart, S; Pfister, X., Caño de Andrade, M.C., Laffitte, J.A. *Tetrahedron: Asymmetry*, 1994, **5**, 665–74. Genêt, J.-P., Pinel, C., Ratovelomanana-Vidal, V., Mallart, S., Pfister, X., Bischoff, L., Caño de Andrade, M.C., Darses, S., Galopin, C., Laffitte, J.A. *Tetrahedron: Asymmetry*, 1994, **5**, 675–90.
17. Gautier, I., Ratovelomanana-Vidal, V., Savignac, P., Genêt, J.P. *Tetrahedron Lett.*, 1996, **37**, 7721–4.
18. Tranchier, J.P., Ratovelomanana-Vidal, V., Genêt, J.P., Tong, S., Cohen, T. *Tetrahedron Lett.*, 1997, **38**, 2951–4.
19. Bertus, P., Phansavath, P., Ratovelomanana-Vidal, V., Genêt, J.P., Touati, R., Homri, T., Ben Hassine, B. *Tetrahedron Lett.*, 1999, **40**, 3175–8. Bertus, P., Phansavath, P., Ratovelomanana-Vidal, V., Genêt, J.P., Touati, R., Homri, T., Ben Hassine, B. *Tetrahedron Asymmetry*, 1999, **10**, 1369–80.
20. Blanc, D., Ratovelomanana-Vidal, V., Marinetti, A., Genêt, J.P. *Synlett*, 1999, **4**, 480–2.
21. Blanc, D., Henry, J.C., Ratovelomanana-Vidal, V., Genêt, J.P. *Tetrahedron Lett.*, 1997, **38**, 6603–6.
22. Marinetti, A., Genêt, J.P., Jus, S., Blanc, D., Ratovelomanana-Vidal, V. *Chem. Eur. J.*, 1999, **5**, 1160–5.
23. Genêt, J.P., Caño de Andrade, M.C., Ratovelomanana-Vidal, V. *Tetrahedron Lett.*, 1995, **36**, 2003–6.
24. Coulon E., Caño de Andrade, M.C., Ratovelomanana-Vidal, V., Genêt, J.P. *Tetrahedron Lett.* 1998, **39**, 6467–70.
25. Review: Ratovelomanana-Vidal, V., Genêt, J.P. *J. Organomet. Chem.* 1998, **567**, 163–71.
26. Muller-Westerhoff, U.T., Yang, Z., Ingram, G. *J. Organomet. Chem.* 1993, **463**, 163.

27. Carroll, M.A., Widdowson, D.A., Williams, D.J. *Synlett*, 1994, 1025.
28. Gokel, G.W., Marquarding, D., Ugi, I.K. *J. Org. Chem.*, 1972, **37**, 3052.
29. (a) Hayashi, T., Mise, T., Fukushima, M., Kagotani, M., Nagashima, N., Hamada, Y., Matsumoto, A., Kawakami, S., Konishi, M., Yamamoto, K., Kumada, M. *Bull. Chem. Soc. Jpn.*, 1980, **53**, 1138. (b) Hayashi, T., Yamazaki, A. *J. Organoment, Chem.*, 1991, **413**, 295.
30. a) Itsuno, S., Ito, K., Hirao, A., Nakahama, S. *J. Chem. Soc., Chem. Commun.*, 1983, 469.
    b) Corey, E.J., Bakshi, R.K., Shibata, S. *J. Am. Chem. Soc.*, 1987, **109**, 5551.
31. Wright, J., Frambes, L., Reeves, P. *J. Organomet. Chem.*, 1994, **476**, 215.
32. Schwink, L., Knochel, P. *Chem. Eur. J.*, 1998, **4**, 950.
33. Almena Perea, J.J., Lotz, M., Knochel, P. *Tetrahedron: Asymmetry*, 1999, **10**, 375.
34. Almena Perea, J.J., Börner, A., Knochel, P. *Tetrahedron Lett.*, 1998, **39**, 8073.
35. Lotz, M., Ireland, T., Almena Perea, J.J., Knochel, P. *Tetrahedron: Asymmetry*, 1999, **10**, 1839.

13 Employment of Catalysts Working in Tandem

CONTENTS

13.1   A ONE-POT SEQUENTIAL ASYMMETRIC HYDROGENATION UTILIZING Rh(I)- AND Ru(II)-CATALYSTS

TAKAYUKI DOI and TAKASHI TAKAHASHI

Department of Applied Chemistry, Graduate School of Science and Engineering, Tokyo Institute of Technology, 2-12-1 Ookayama, Meguro, Tokyo 152–8552, JAPAN TEL +(81) 3-5734–2120, FAX +(81) 3-5734–2884, Email: ttakashi@o.cc.titech.ac.jp, doit@mep.titech.ac.jp.

Asymmetric hydrogenation of enamido β-keto esters was carried out in the presence of both Rh(I)- and Ru(II)-chiral phosphine complexes as catalysts[1]. This is the efficient method to prepare statin analogues. The process independently induces two stereo centres in a molecule in a simple manner.

13.1.1   SYNTHESIS OF ETHYL (Z)-4-ACETAMIDO-3-OXO-5-PHENYL-4-PENTENOATE[1]

Materials and equipment

- (Z)-2-Acetamino-cinnamic acid[2,3], 1.03 g, 5.0 mmol
- Dry tetrahydrofuran, 20 mL
- 1,1'-Carbonyldiimidazole, 1.28 g, 5.5 mmol
- Lithiated ethyl acetate, 15 mmol
- Saturated aqueous ammonium chloride solution, 20 mL

- Saturated aqueous sodium bicarbonate solution, 30 mL
- Brine, 30 mL
- Magnesium sulfate, 3 g
- Ethyl acetate, hexane
- Silica gel 60
- Column chromatography
- Test tubes, 25 mL × 50

- 100 mL Three-necked and 50 mL two-necked round-bottom flask with magnetic stirrer bars
- Magnetic stirrer
- Dry ice–acetone cooling bath
- Thermometer, $-80\,^\circ$C to $30\,^\circ$C
- Syringes
- TLC
- Separatory funnel
- Cannula
- Rotary evaporator

Procedure[4]

1. To (Z)-2-acetamino-cinnamic acid (1.03 g, 5.0 mmol) in dry THF (20 mL) was added 1,1'-carbonyldiimidazole (1.28 g, 5.5 mmol) at room temperature; then lithiated ethyl acetate (15 mmol) was added via cannula at $-78\,^\circ$C.
2. The reaction mixture was stirred at the same temperature for 30 min, then stirred at $0\,^\circ$C for 30 min, and poured into NH_4Cl aq. solution.
3. The aqueous layer was extracted with ethyl acetate and combined organic layer was washed with $NaHCO_3$ and brine, then dried over $MgSO_4$.
4. After removal of the solvent, the residue was chromatographed over silica gel to afford ethyl (Z)-4-acetamido-3-oxo-5-phenyl-4-pentenoate (870 mg, 63 %).

13.1.2 ASYMMETRIC HYDROGENATION OF ETHYL 4-ACETAMIDO-3-OXO-5-PHENYL-4-PENTENOATE

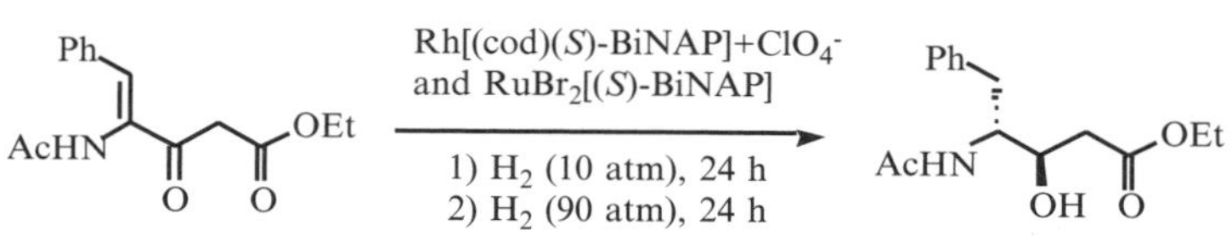

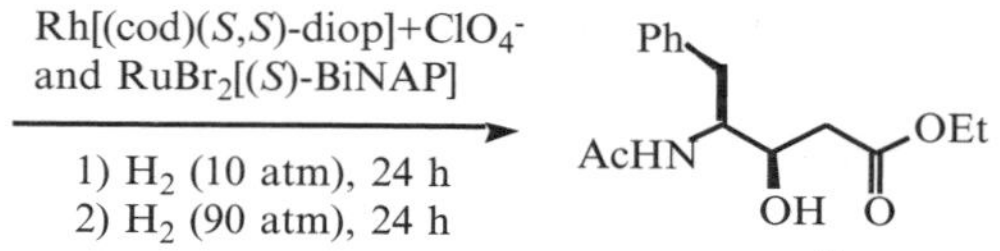

Materials and equipment

- Ethyl (Z)-4-acetamido-3-oxo-5-phenyl-4-pentenoate, 275 mg, 1.0 mmol
- $[Rh(cod)(S)\text{-BiNAP}]^+CiO_4^-$ [5], 9.3 mg, 0.01 mmol
- $RuBr_2(S)\text{-BiNAP}$ [6,7], 8.8 mg, 0.01 mmol
- Distilled triethylamine, 0.14 mL, 1.0 mmol
- Distilled ethanol, 10 mL
- 1 N HCl solution, 20 mL
- Brine, 30 mL
- Ethyl acetate, hexane
- Silica gel 60

- 50 mL Autoclave with glasstube and a magnetic stirrer bar
- Hydrogen gas tank
- Gas connector from tank to autoclave
- Magnetic stirrer
- Oil-bath
- Thermometer
- Syringes
- TLC
- Rotary evaporator

Procedure

1. In a 50 mL autoclave containing a glass tube and magnetic stirrer bar were placed $[Rh(cod)(S)\text{-BiNAP}]^+CiO_4^-$ and $RuBr_2[(S)\text{-BiNAP}]$ as catalysts.
2. To this mixture were added the substrate **1**, triethylamine, and ethanol. The autoclave was filled with hydrogen (10 atm) after repeated (4–5 times) filling and purging of hydrogen.
3. The reaction was carried out under 10 atm H_2 at 40 °C for 24 h and under 90 atm at 40 °C for an additional 24 hours.
4. The solvent was removed under reduced pressure.
5. The residue was diluted with ethyl acetate before being washed with 1 N HCl at 0 °C. The aqueous layer was extracted with ethyl acetate and the combined organic layer was washed with brine and dried over Na_2SO_4.
6. After removal of the solvent, the residue was eluted through a short silica gel column to remove the catalyst (elution with hexane:ethyl acetate = 1:2). The eluent was concentrated *in vacuo* to give the product **2** (99 % yield) and the diastereoselectivity was determined by HPLC analysis (99 %). The enantioselectivity of the product was determined by ^1H NMR analysis with chiral shift reagent (+)-Eu(dppm) in $CDCl_3$ and by chiral HPLC analysis (Chiralcel-OD).

Ethyl (3R,4R)-4-(acetamido)-3-hydroxy-5-phenylpentanoate (2)

^{1}H NMR (270 MHz, CDCl$_3$): δ 1.28 (3H, t, $J = 6.9$ Hz), 1.99 (3H, s), 2.38 (2H, ABX, $J = 3.0$, 17.3 Hz), 2.57 (1H, ABX, $J = 10.3$, 17.3 Hz), 2.91 (2H, d, $J = 7.6$ Hz), 3.60–4.05 (2H, m), 4.15 (2H, q, $J = 6.9$ Hz), 5.83–5.87 (1H, m), 7.19–7.49 (5H, m, aromatic).

^{13}C NMR (67.5 MHz, CDCl$_3$). δ 173.5, 170.0, 137.9, 129.3, 128.6, 126.5, 66.8, 60.9, 53.9, 38.6, 38.2, 23.4, 14.1.

IR(CHCl$_3$) 3350, 2922, 1729, 1647, 1537, 1372, 1295 cm^{-1}; MS (EI, 70 eV) 279 (M$^+$, 11%), 220 (9), 192 (14), 167 (19), 149 (68), 135 (49).

HRMS (EI, 70 eV) calculated for C$_{14}$H$_{15}$NO$_4$ 279.1471 (M$^+$), found 279.1434.

HPLC (Silica gel 60–5 mm, 7.5 o.d.x 300 mm, elution with 12% 2-propanol in hexane, 3.0 mL/min) Rt 29–33 min; [α]$_D^{25}$ = +69.0° ($c = 0.116$, MeOH) (>95% ee).

Use of [Rh(cod)(*S, S*)-diop]$^+$CiO$_4^{-[8]}$ instead of [Rh(cod)(*S*)-BINAP]$^+$ CiO$_4^-$ gave its diastereomer **3** with 72% stereoselectivity and >95% ee.

Ethyl (3R,4S)-4-(acetamido)-3-hydroxy-5-phenylpentanoate (3)

^{1}H NMR (270 MHz, CDCl$_3$): δ 1.28 (3H, t, $J = 6.9$ Hz), 1.88 (3H, s), 2.53 (3H, m), 2.87 (1H, ABX, $J = 8.6$, 14.0 Hz), 2.99 (1H, ABX, $J = 4.6$, 14.0 Hz), 3.76–4.15 (2H, m), 4.20 (2H, q, $J = 6.9$ Hz), 5.53–5.49 (1H, m), 7.19–7.53 (5H, m, aromatic).

^{13}C NMR (67.5 MHz, CDCl$_3$): δ 172.9, 169.1, 141.5, 129.3, 128.6, 126.7, 68.9, 60.9, 54.2, 38.2, 35.1, 23.3, 14.1.

IR (CHCl$_3$) 3350, 2922, 1729, 1647, 1537, 1372, 1295 cm^{-1}.

MS (EI, 70 eV) 279 (M$^+$), 220, 192, 174, 163, 135.

HRMS (EI, 70 eV) calculated for C$_{14}$H$_{15}$NO$_4$ 261.1471 (M$^+$), found 261.1495.

HPLC (Silica gel 60–5 mm, 7.5 o.d.x 300 mm, elution with 12% 2-propanol in hexane, 3.0 mL/min) Rt 34–40 min; [α]$^{25}_D$ = −55.5° ($c = 0.072$, MeOH) (>95% ee).

Conclusion

A direct method for the respective preparation of the core units of statin analogues (3*R*,4*R*)-**2**, (3*S*,4*S*)-**2**, (3*R*,4*S*)-**3**, and (3*S*,4*R*)-**3** in enantiomerically pure form is described. These analogues are prepared from the same molecule **1** in a one-pot, sequential asymmetric hydrogenation process utilizing Rh(I)- and Ru(II)-chiral phosphine complexes. Some other examples are depicted in Table 13.1.

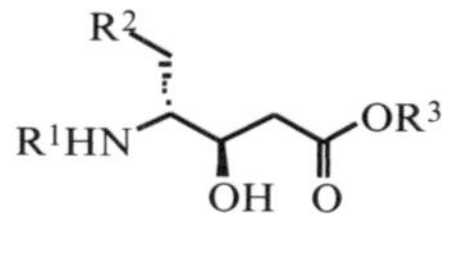

$$R^1HN\overset{\overset{\displaystyle R^2}{|}}{\underset{\underset{\displaystyle OH}{|}}{C}}\overset{O}{\underset{||}{C}}OR^3$$

Table 13.1   Sequential asymmetric hydrogenation of γ-(acylamino)-γ, δ-unsaturated-β-keto esters catalysed by $Rh[(cod)(S)\text{-}BiNAP]^+ClO_4^-$ and $RuBr_2[(S)\text{-}BiNAP]$.

Solvent	R^1	R^2	R^3	Yield %	ee %
EtOH	Ac	Ph	Et	99	>95
MeOH	Ac	Ph	Me	99	>95
t-BuOH	Ac	Ph	t-Bu	no reaction	–
Et	Boc	Ph	Et	99	>95
Et	Ac	4-Cl-C_6H_4	Et	90	>95

REFERENCES

1. Doi, T., Kokubo, M., Yamamoto, K., Takahashi, T. *J. Org. Chem.*, 1998, **63**, 428.
2. *Org. Syn.* II, 1.
3. Carlström, A.-S., Frejd, T. *Synthesis*, 1989, 414.
4. Rich, D.H., Sun, E.T., Boparai, A.S. *J. Org. Chem.*, 1978, **43**, 3624.
5. Miyashita, A., Yasuda, A., Takaya, H., Toriumi, K., Ito, T., Souchi, T., Noyori, R. *J. Am. Chem. Soc.*, 1980, **102**, 7932.
6. Noyori, R., Ohkuma, T., Kitamura, M., Takaya, H., Sayo, N., Kumobayashi, H., Akutagawa, S. *J. Am. Chem. Soc.*, 1987, **109**, 5856.
7. Genêt, J.P., Pinel, C., Ratovelomanana-Vidal, V., Mallart, S., Pfister, X., Cano De Andrade, M.C., Laffitte, J.A. *Tetrahedron: Asymmetry*, 1994, **5**, 665.
8. Sinou, D., Kagan, H.B. *J. Organomet. Chem.*, 1976, **114**, 325.

Index